ファミリーペット

プチしば
petit shiba

PHOTO BY YONEO MORITA

永岡書店

小さな小さなプチしば。

もしかしたら、
あなたのそばにも
いるかもしれません……。

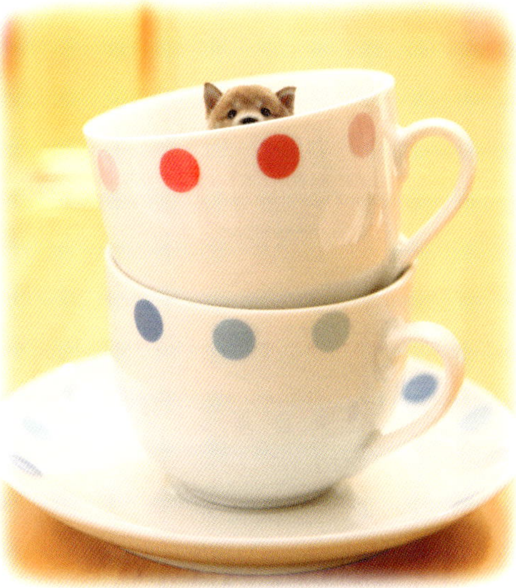

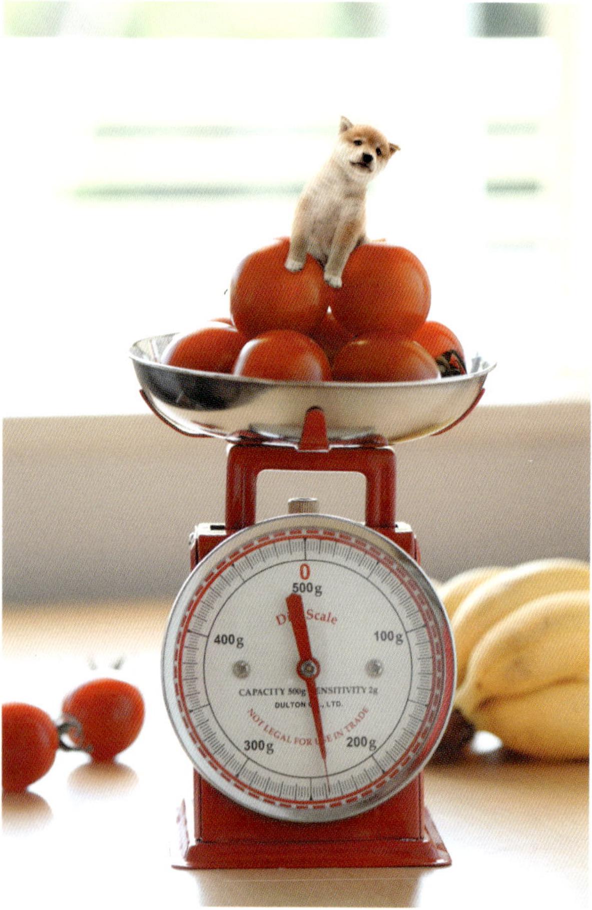
0
500g
400g
100g
300g
200g

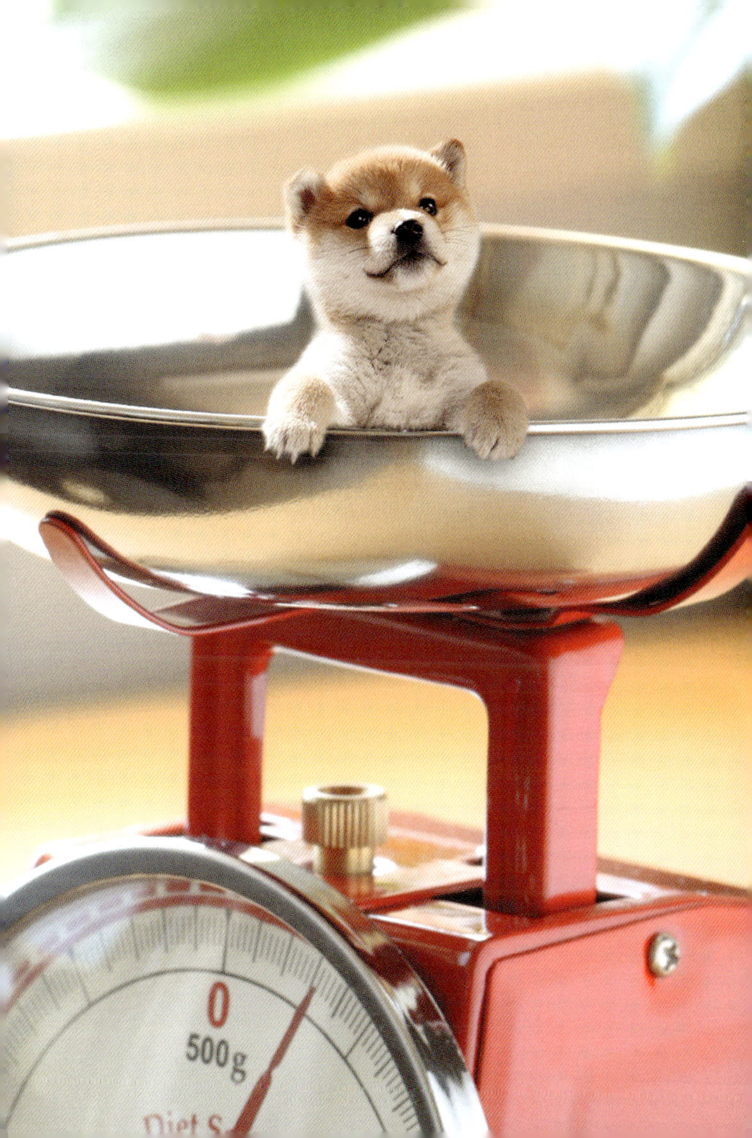
0
500g

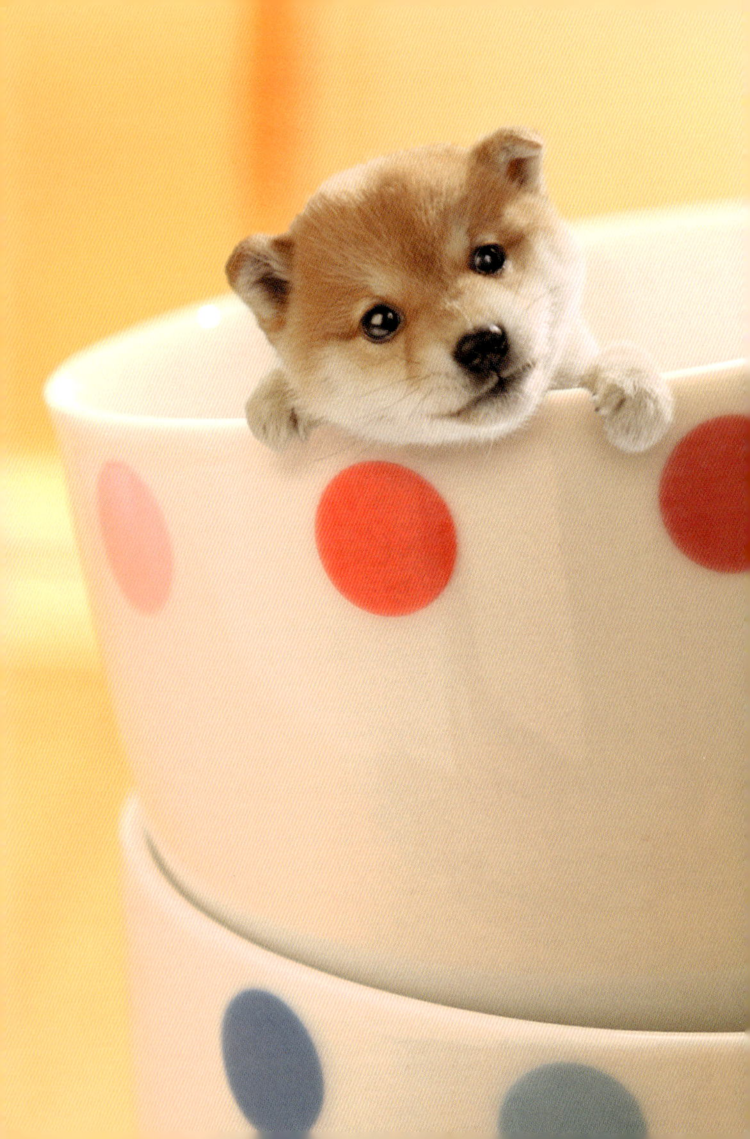

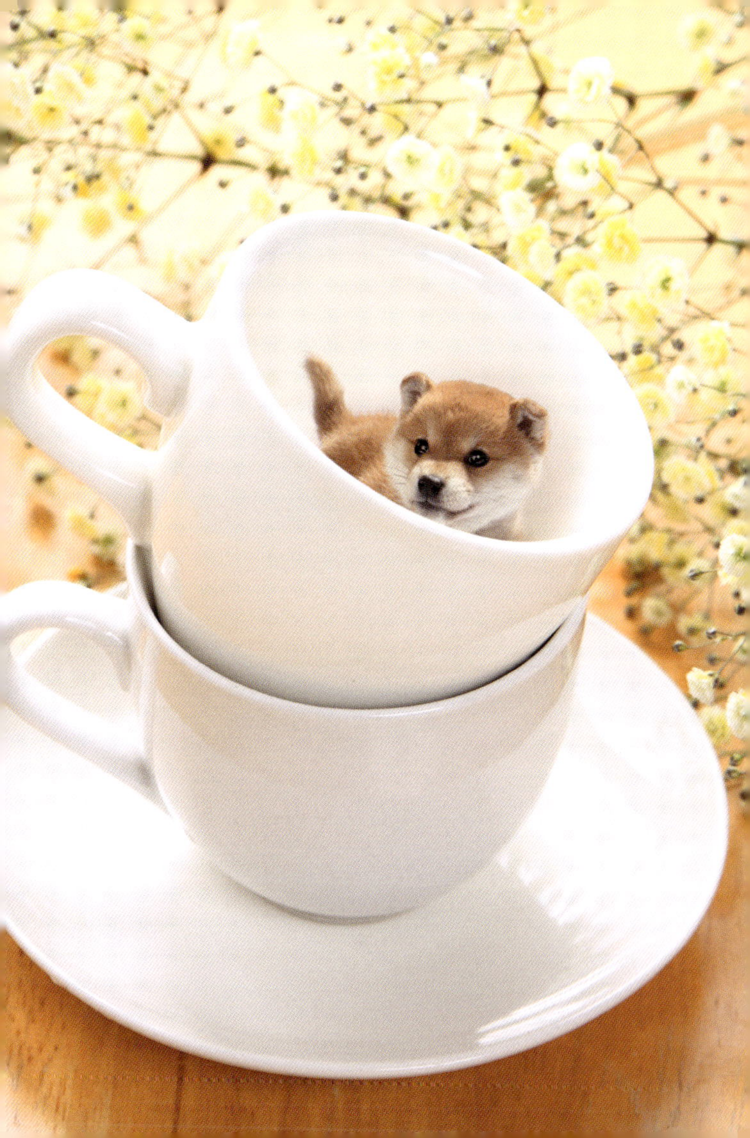

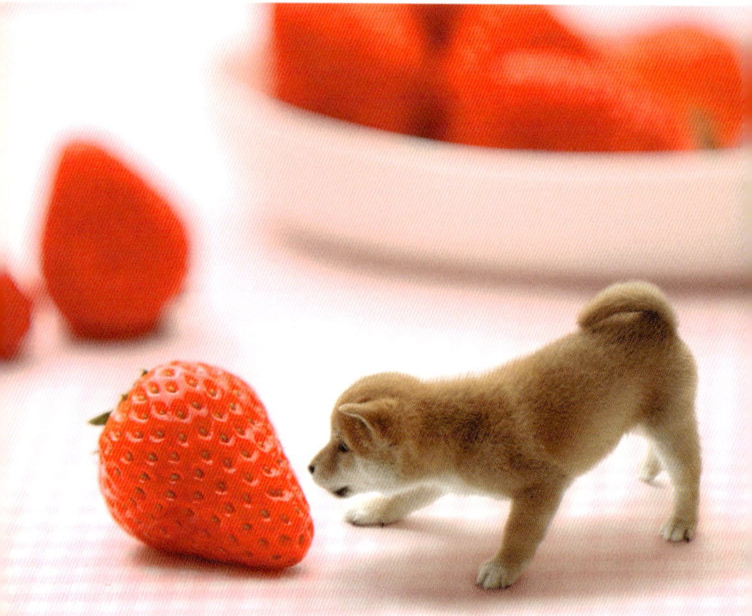

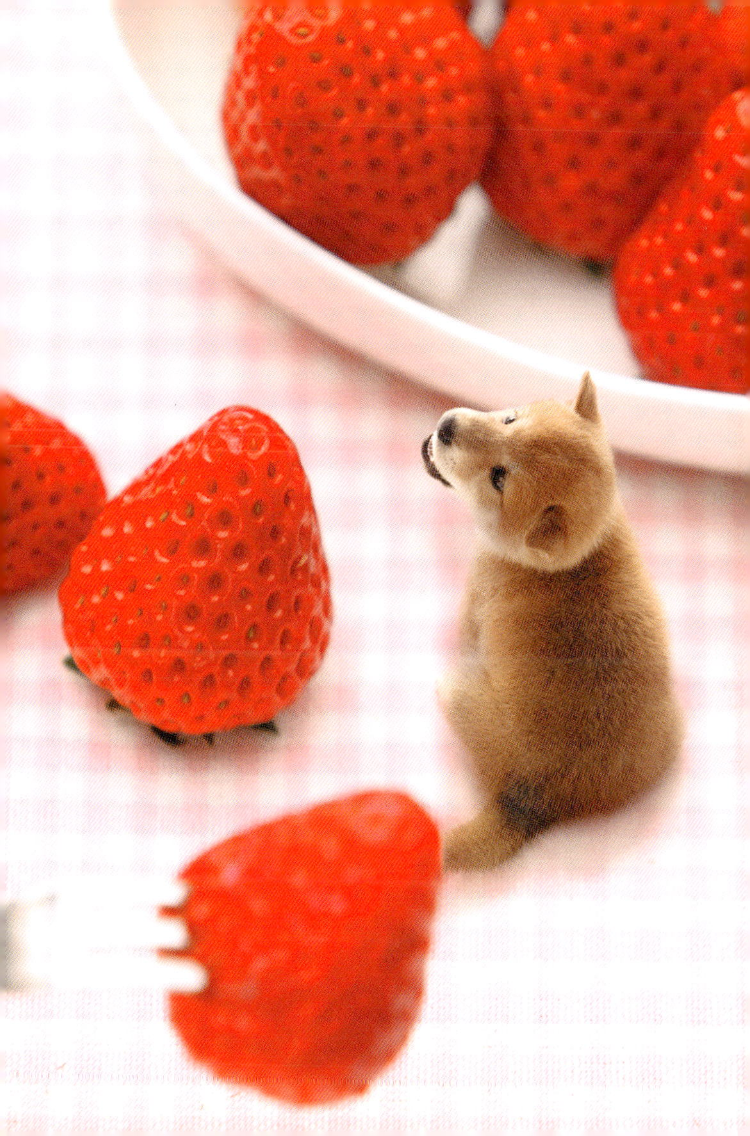

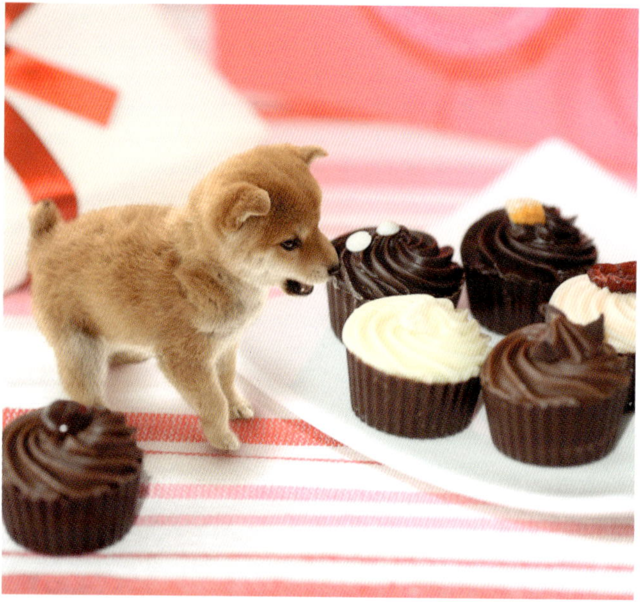

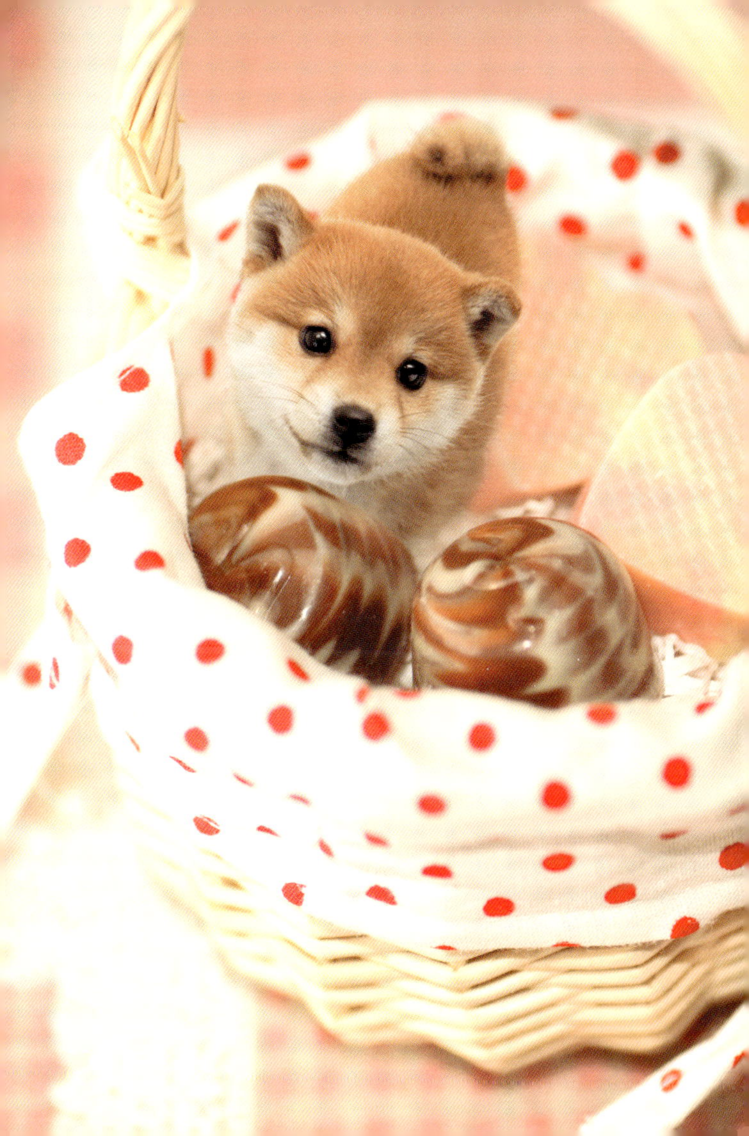

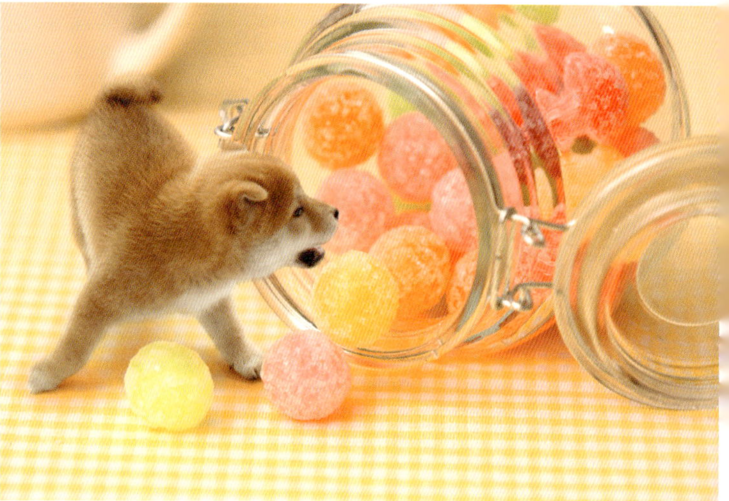

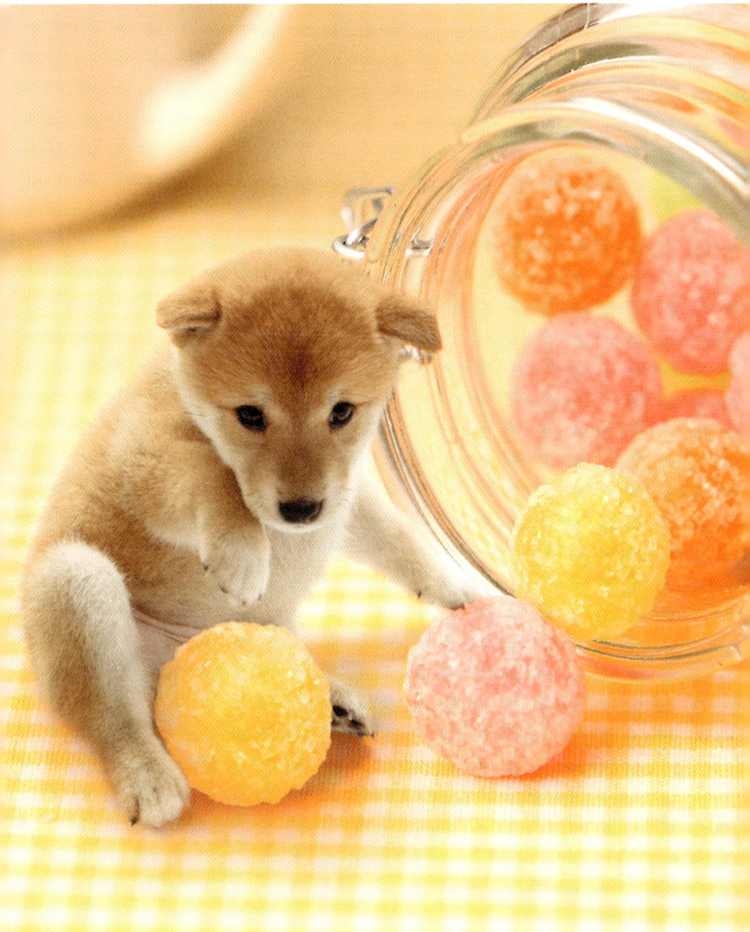

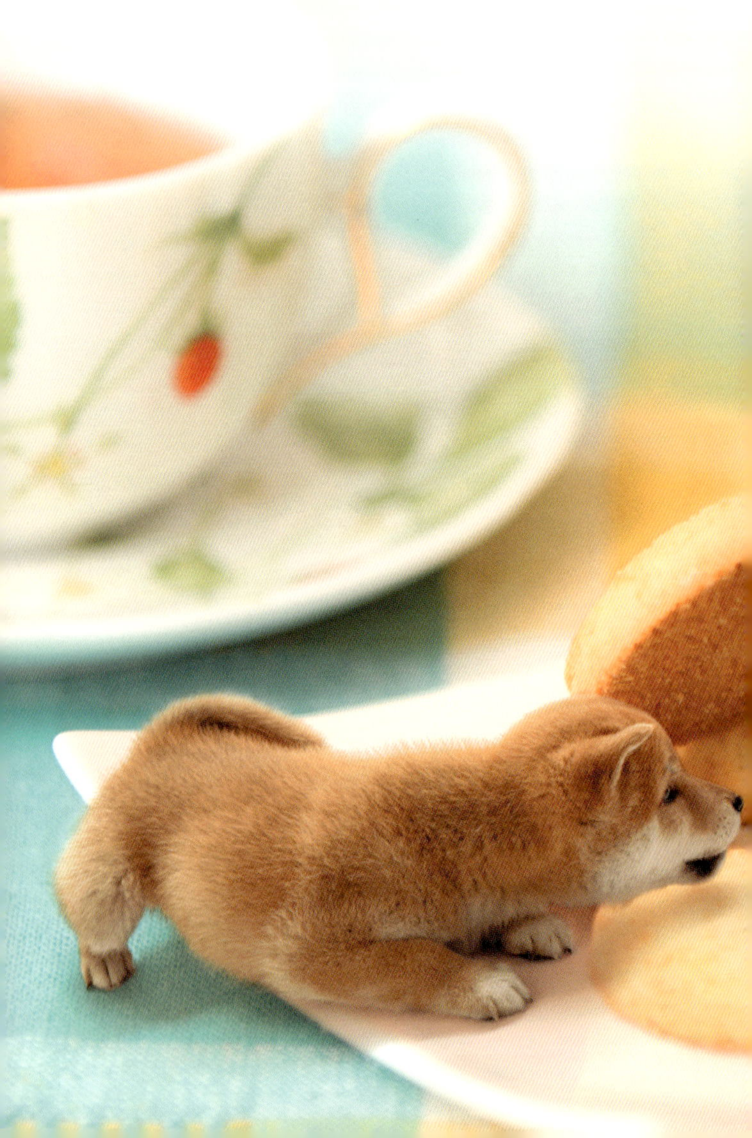

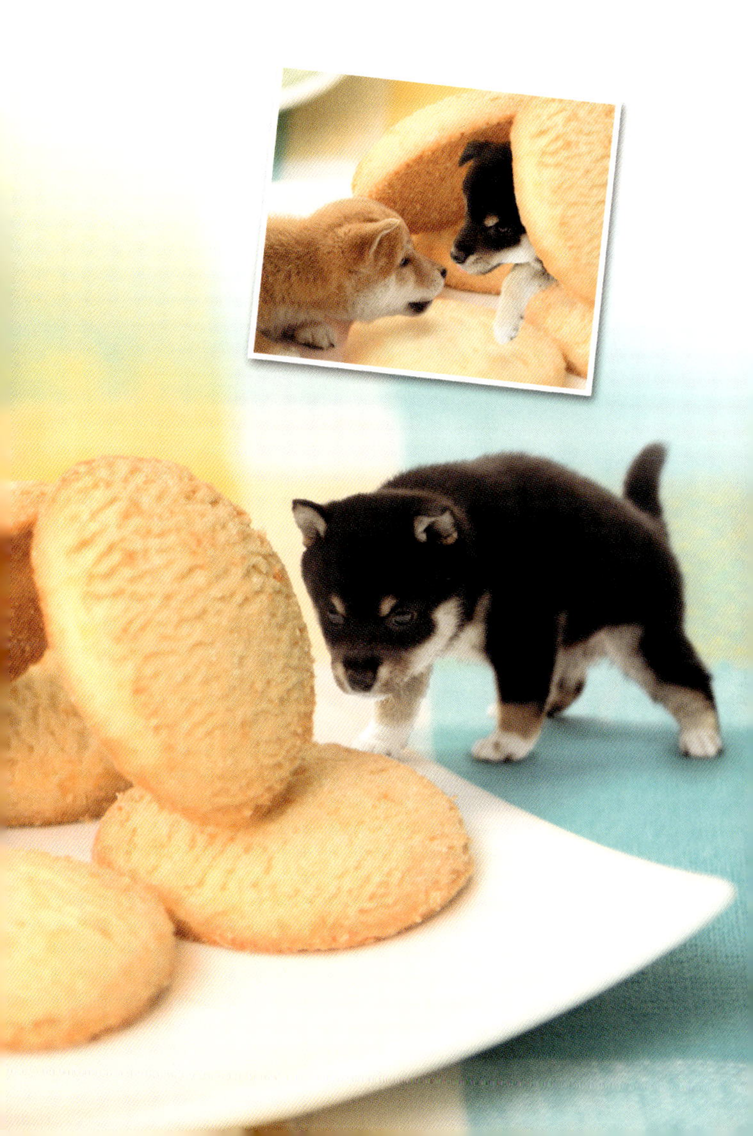

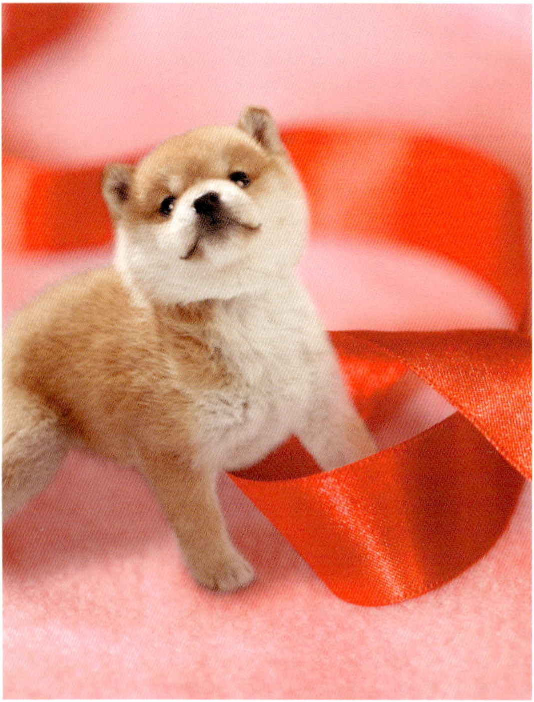

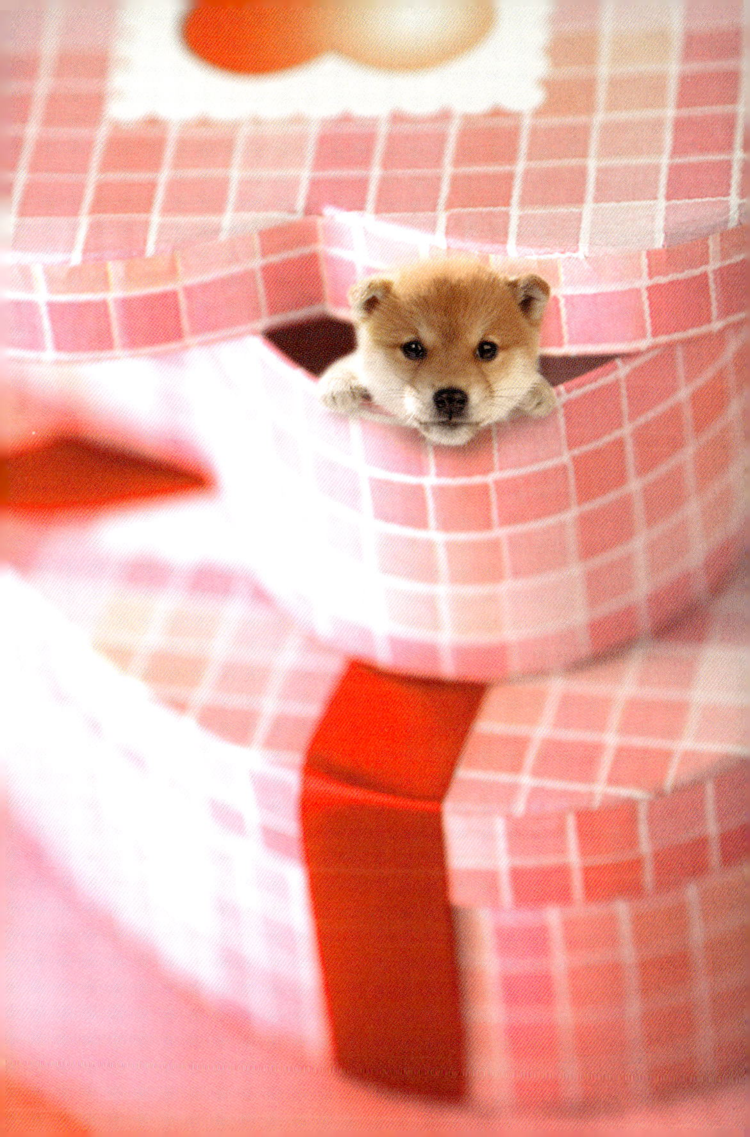

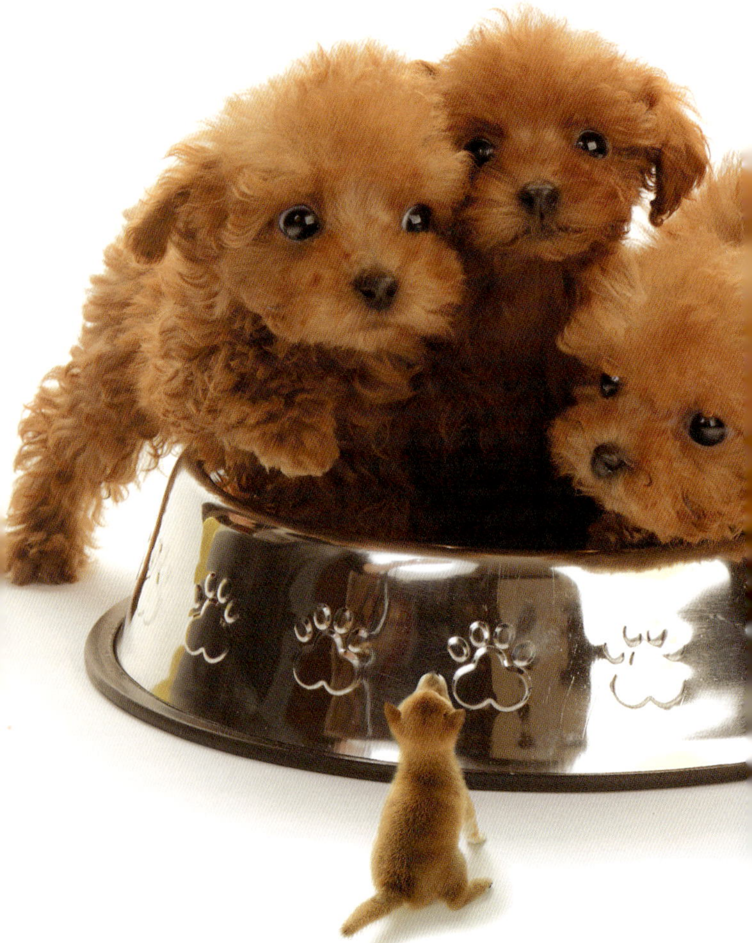

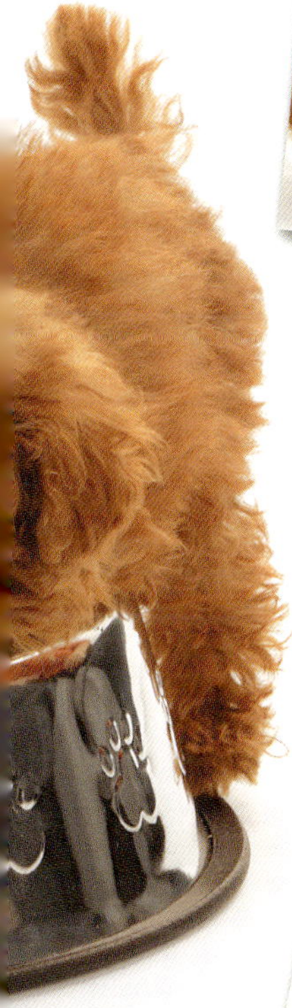
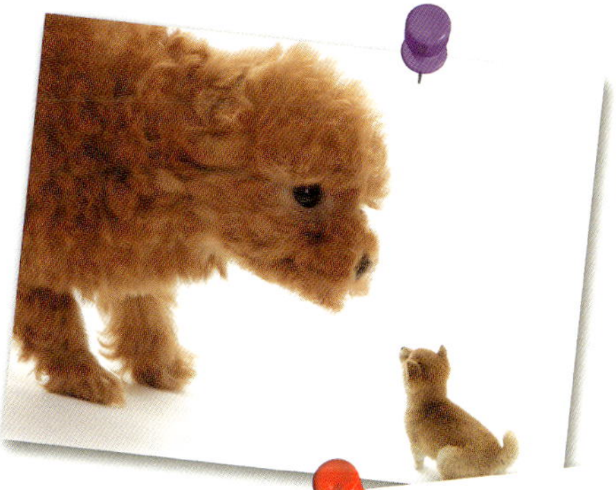
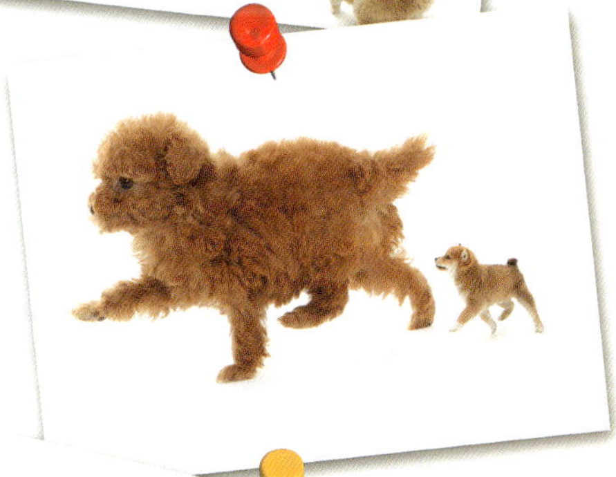
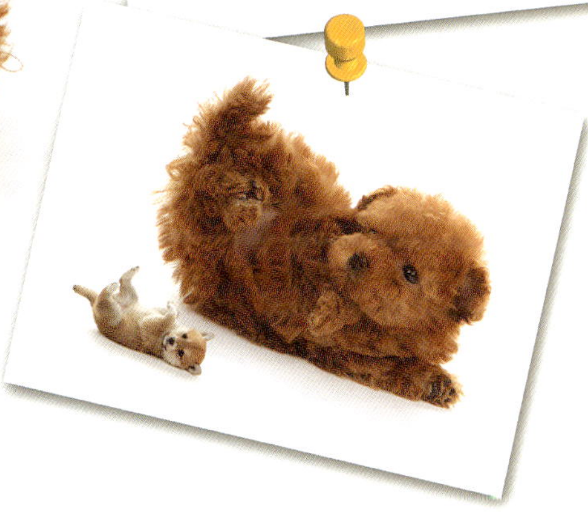

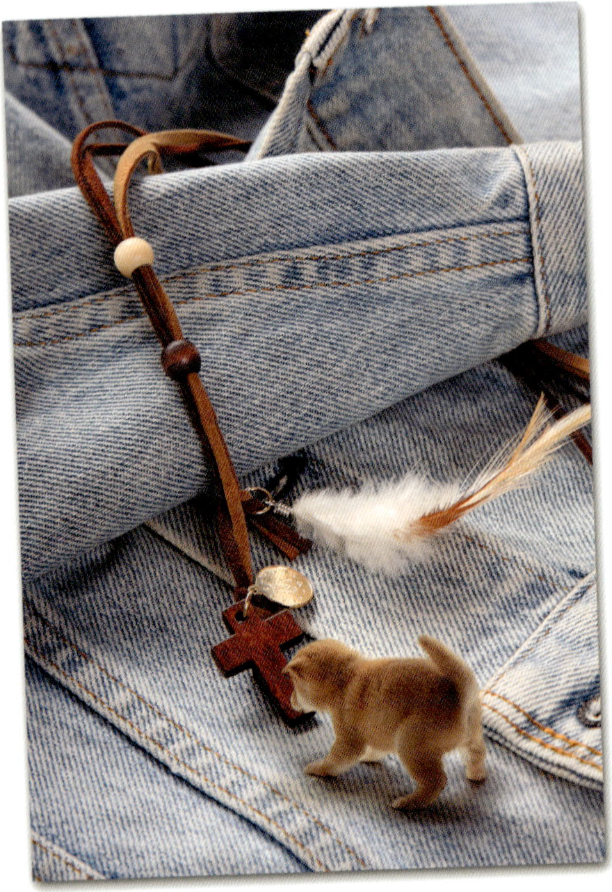

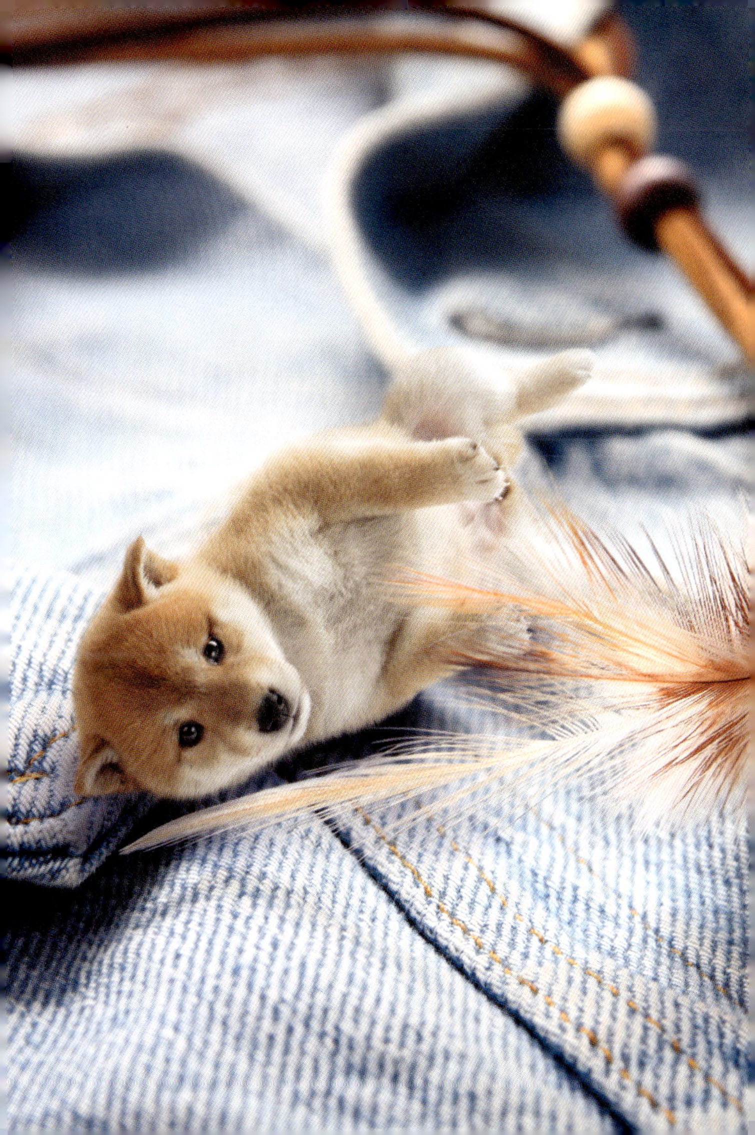

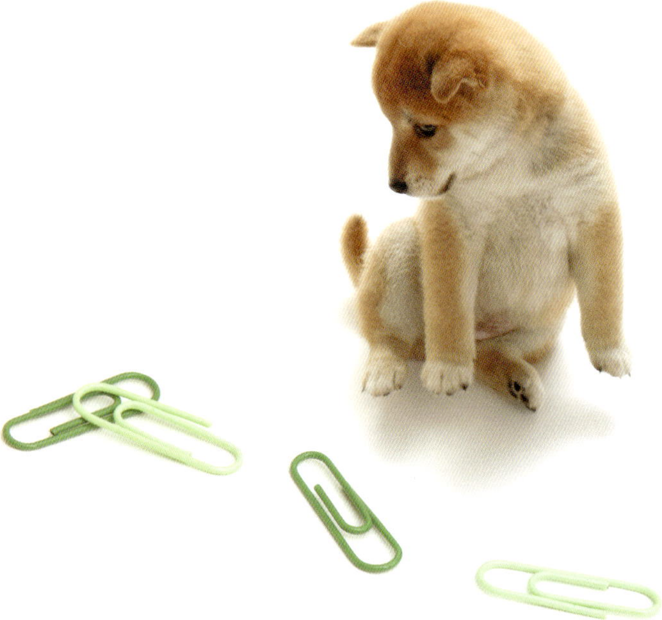

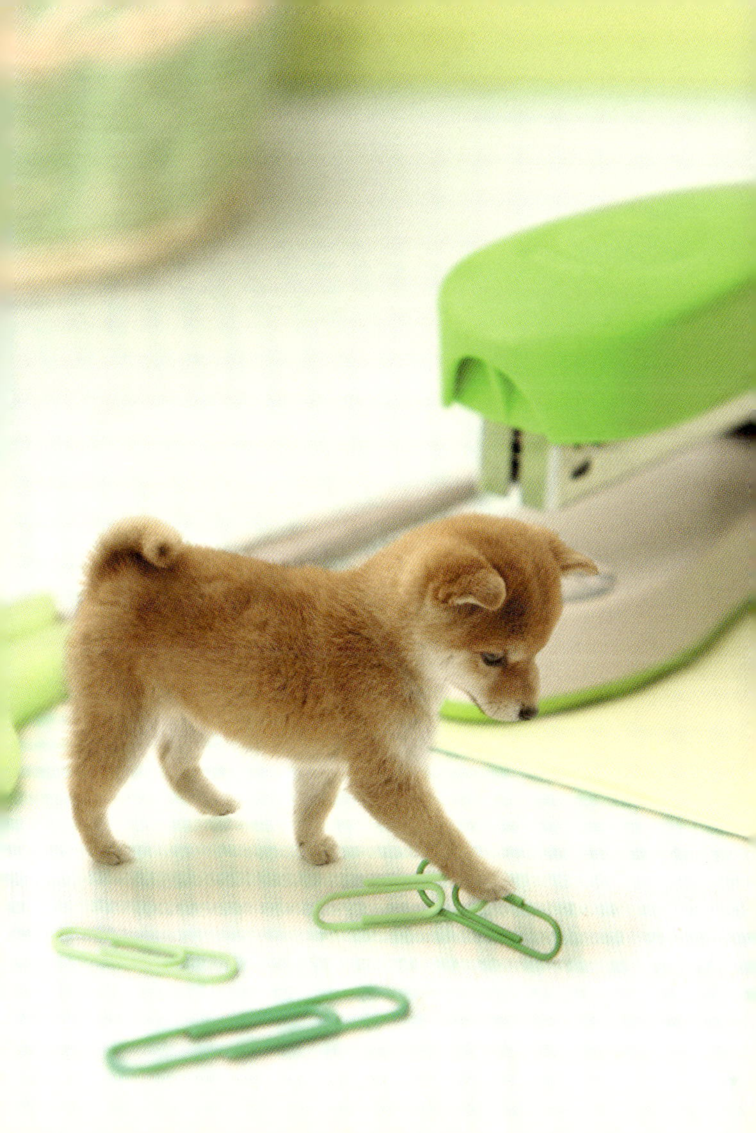

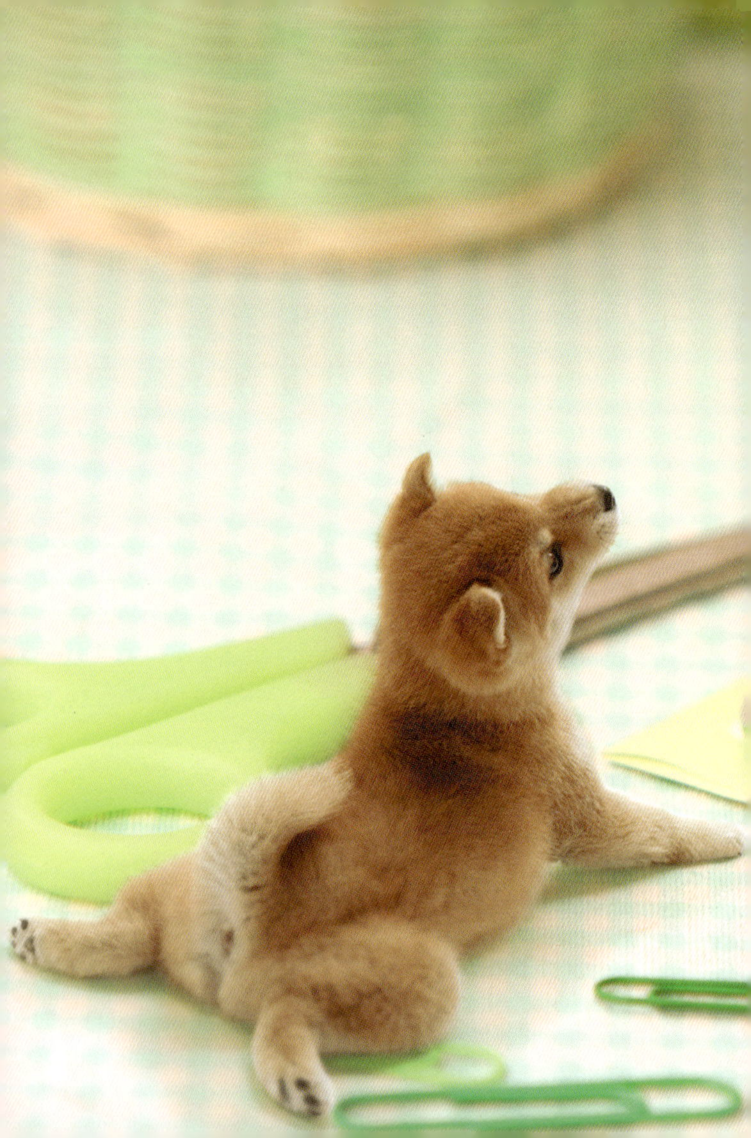

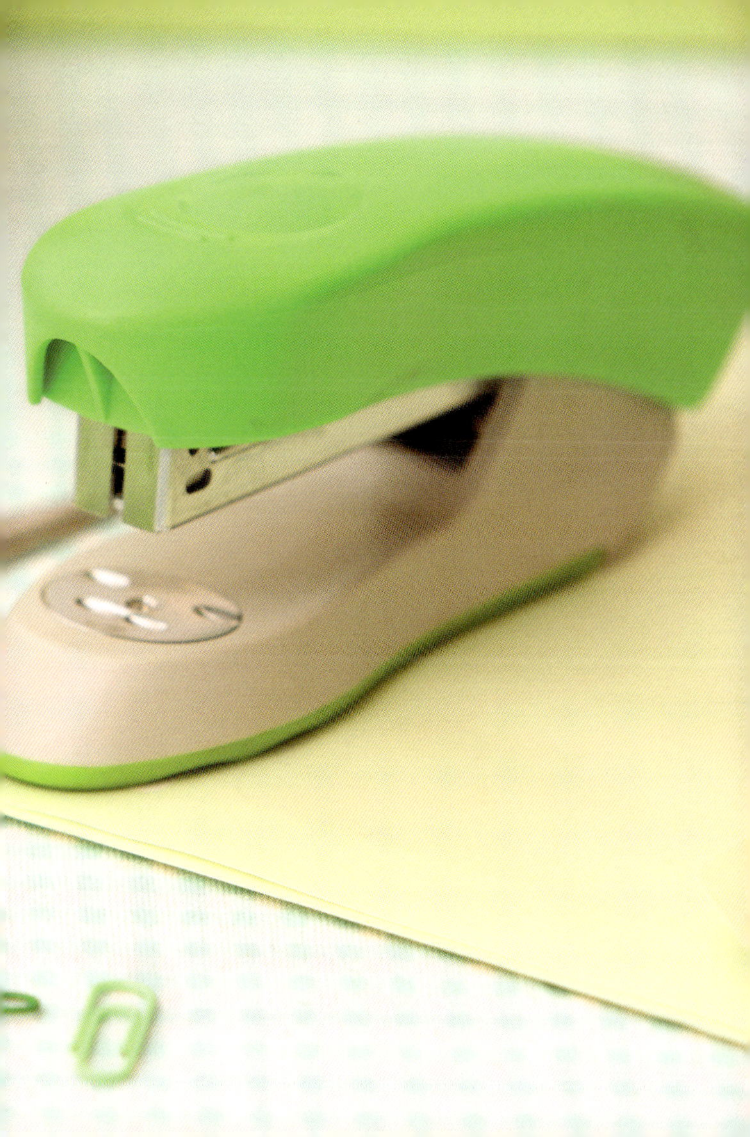

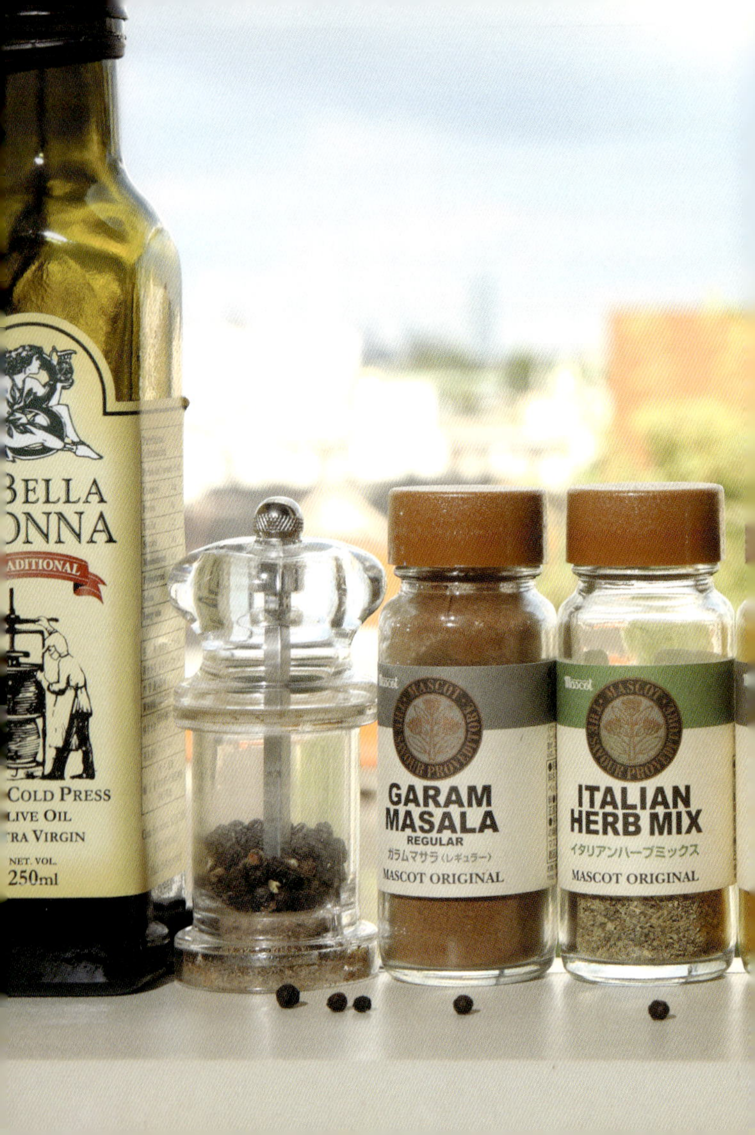
BELLA
ONNA
ADITIONAL
COLD PRESS
LIVE OIL
TRA VIRGIN
NET. VOL.
250ml
GARAM
MASALA
REGULAR
ガラムマサラ〈レギュラー〉
MASCOT ORIGINAL
ITALIAN
HERB MIX
イタリアンハーブミックス
MASCOT ORIGINAL

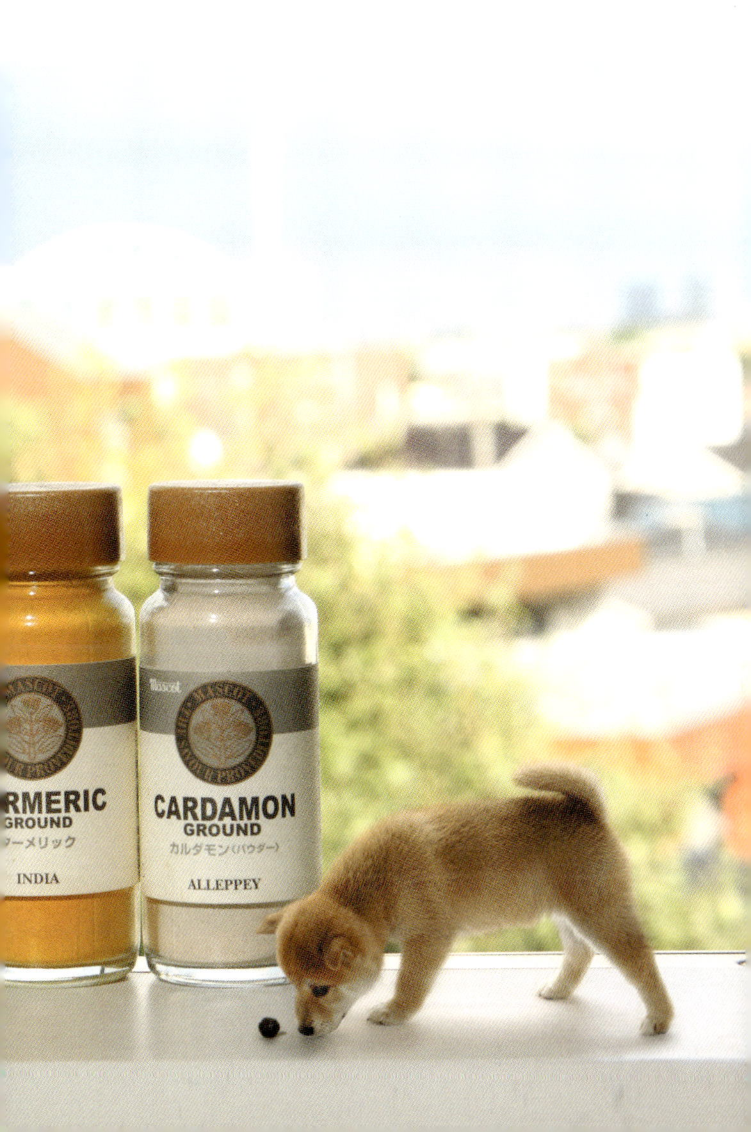
RMERIC
GROUND
ーメリック
INDIA
CARDAMON
GROUND
カルダモン(パウダー)
ALLEPPEY

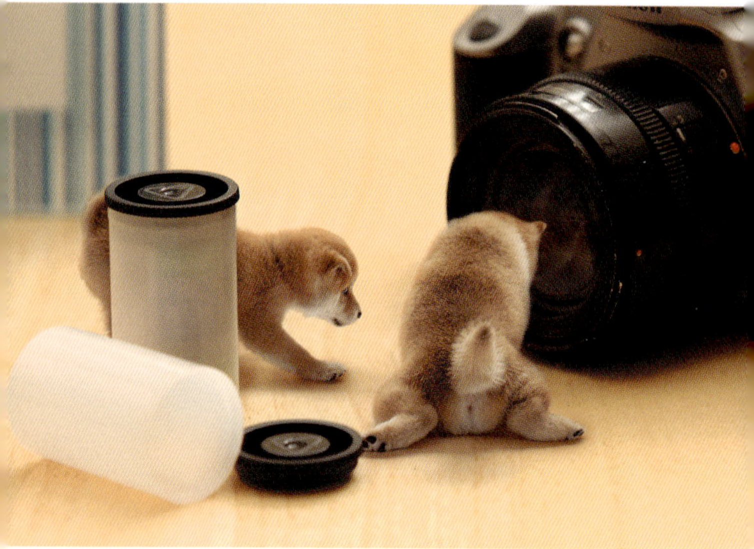

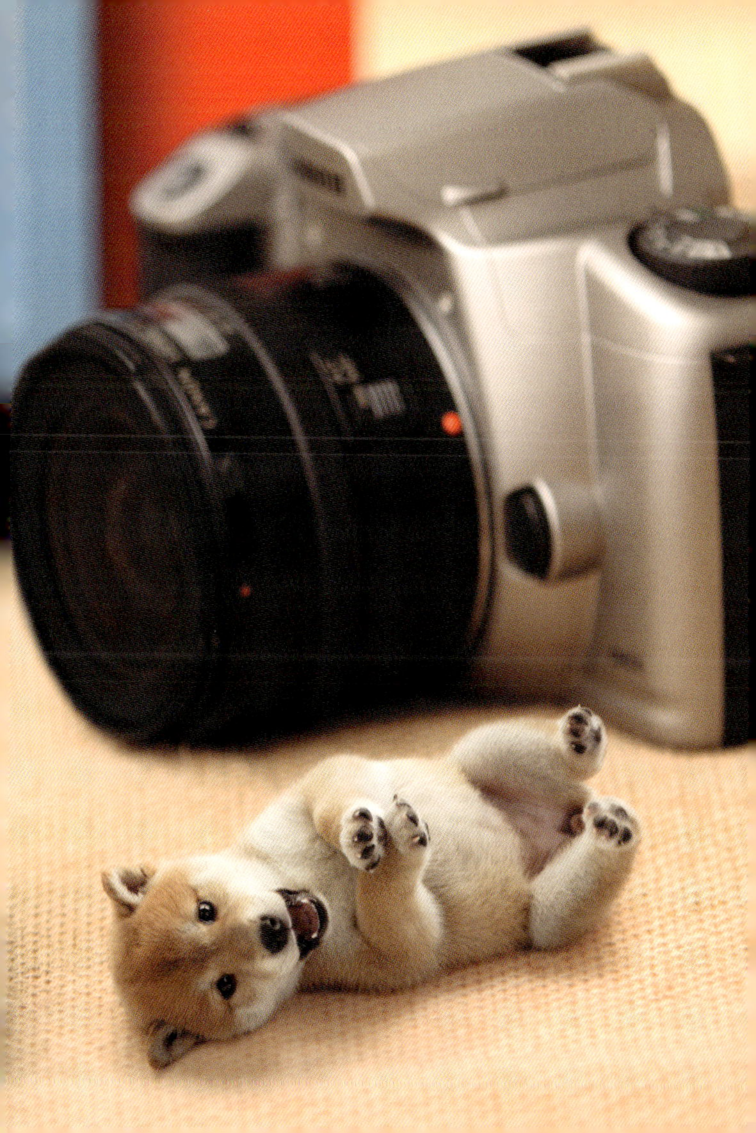

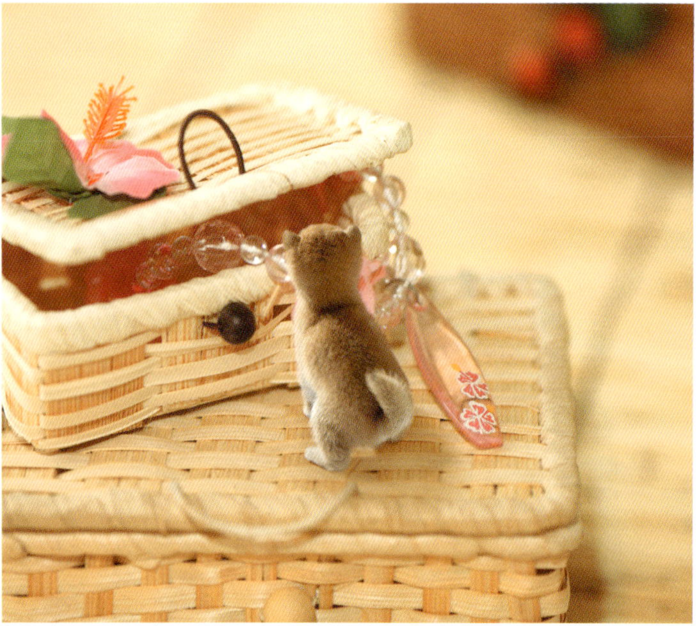

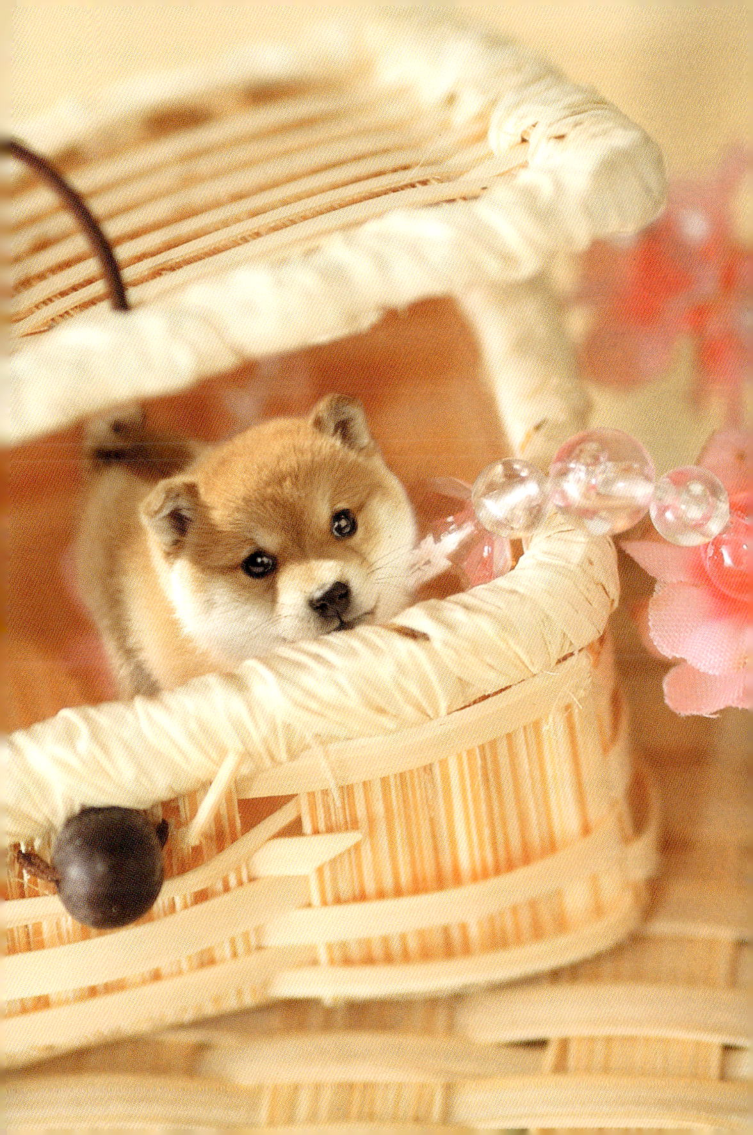

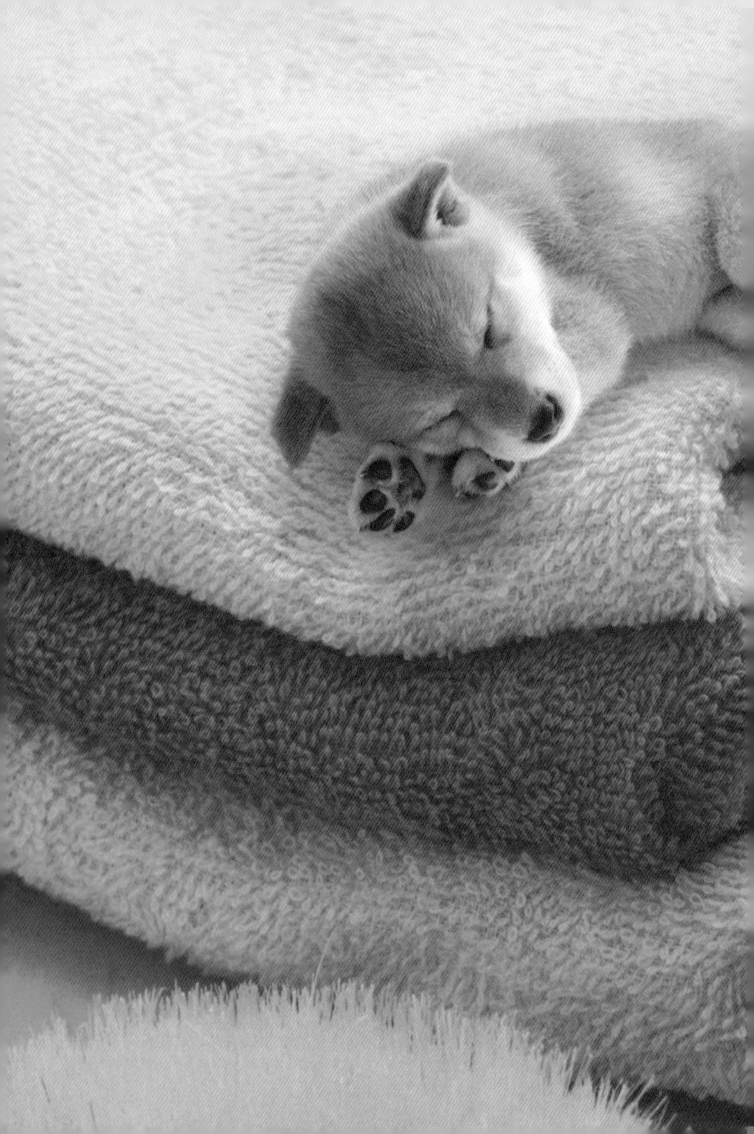

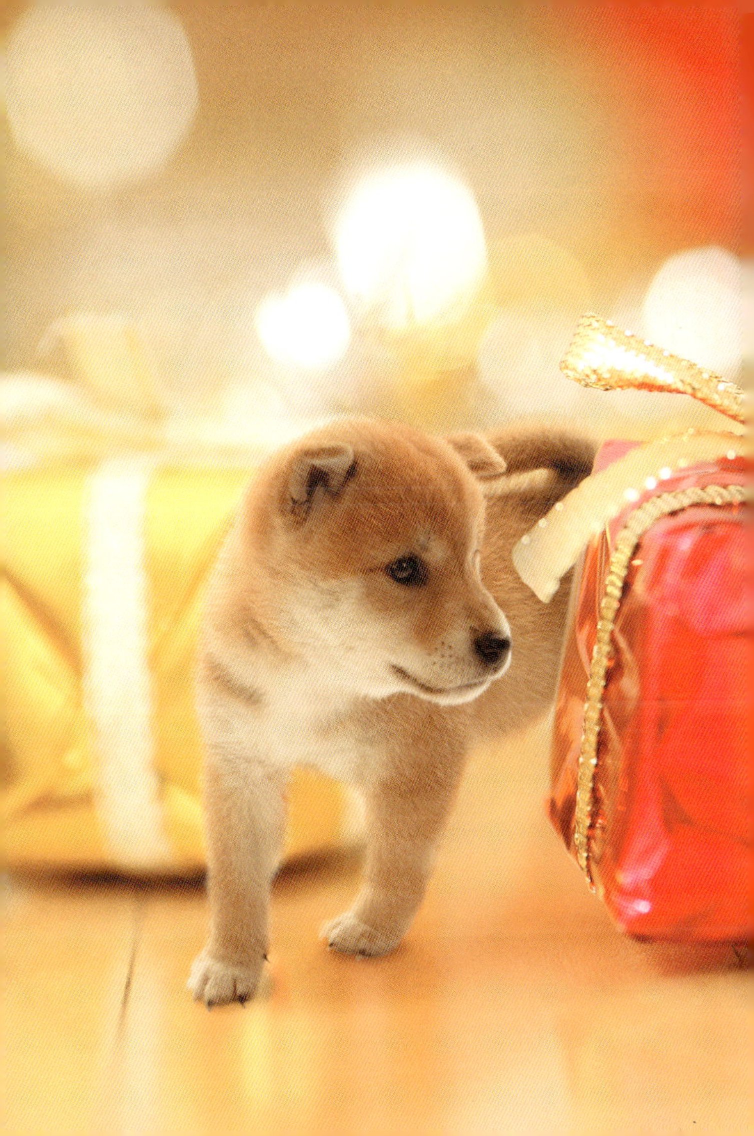

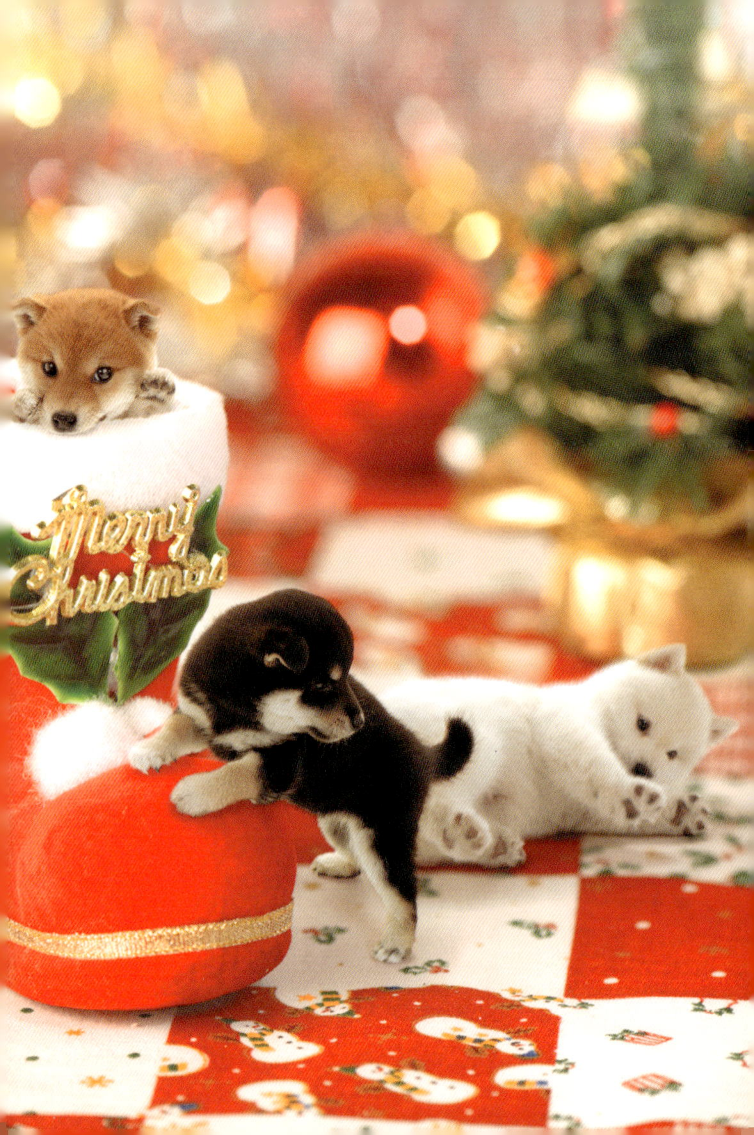
Merry
Christmas

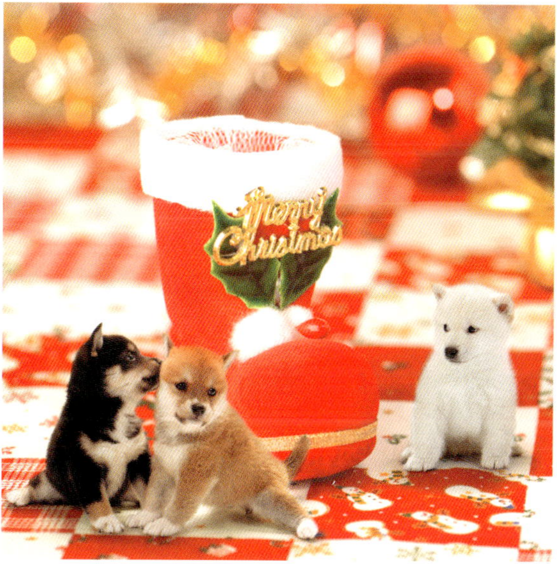
Merry
Christmas

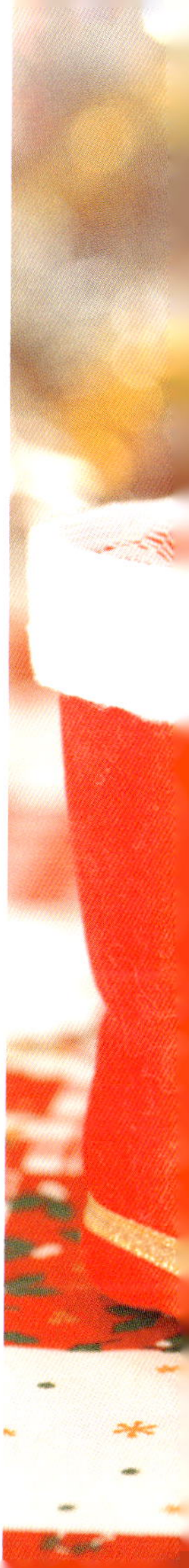

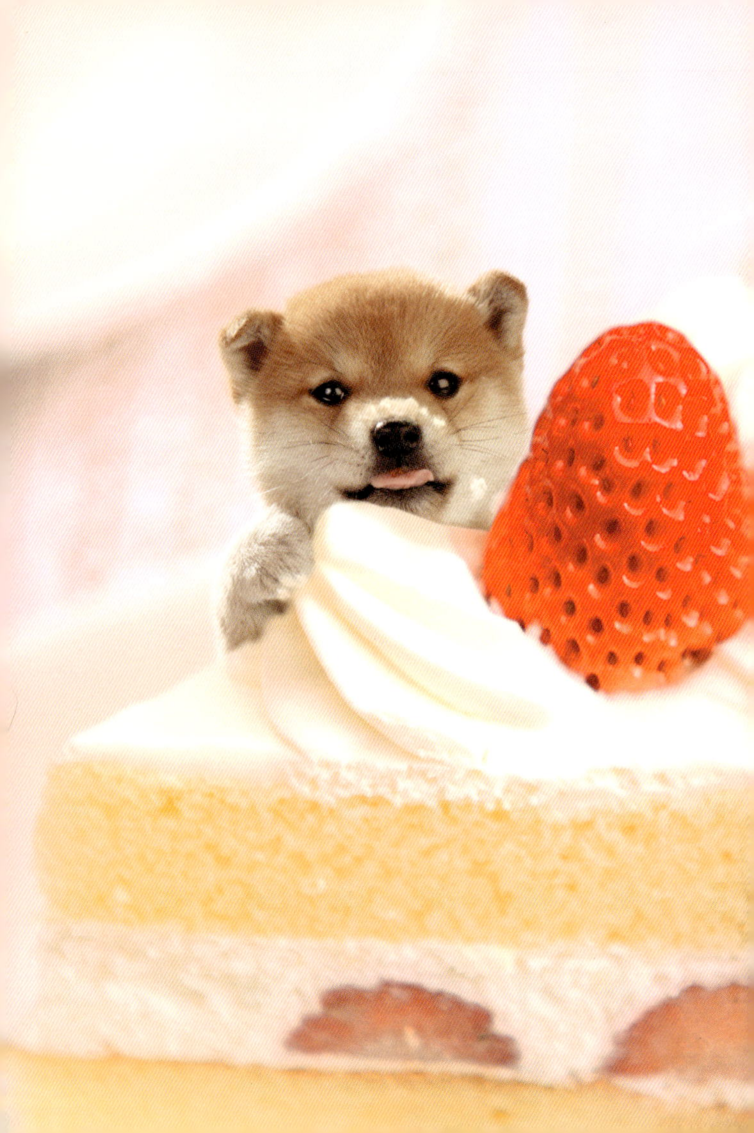

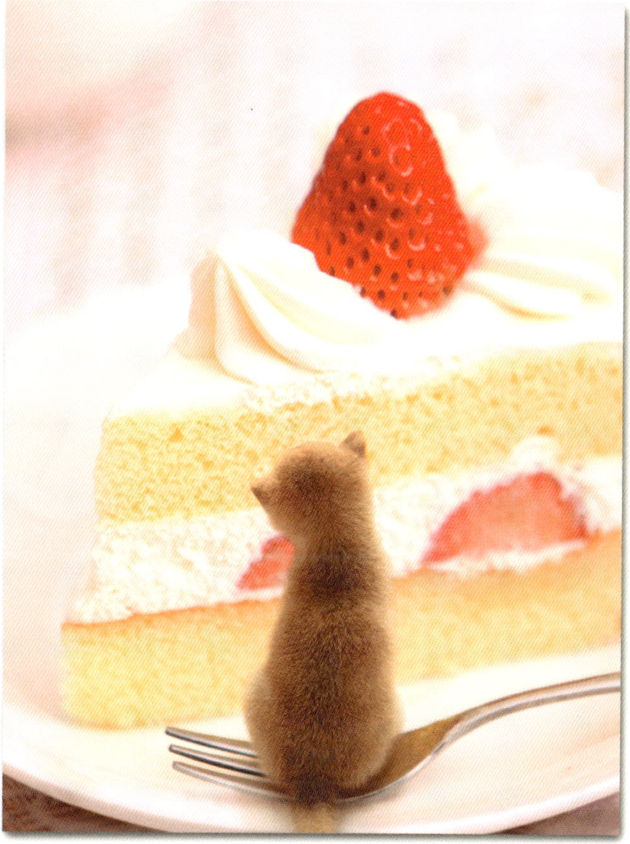

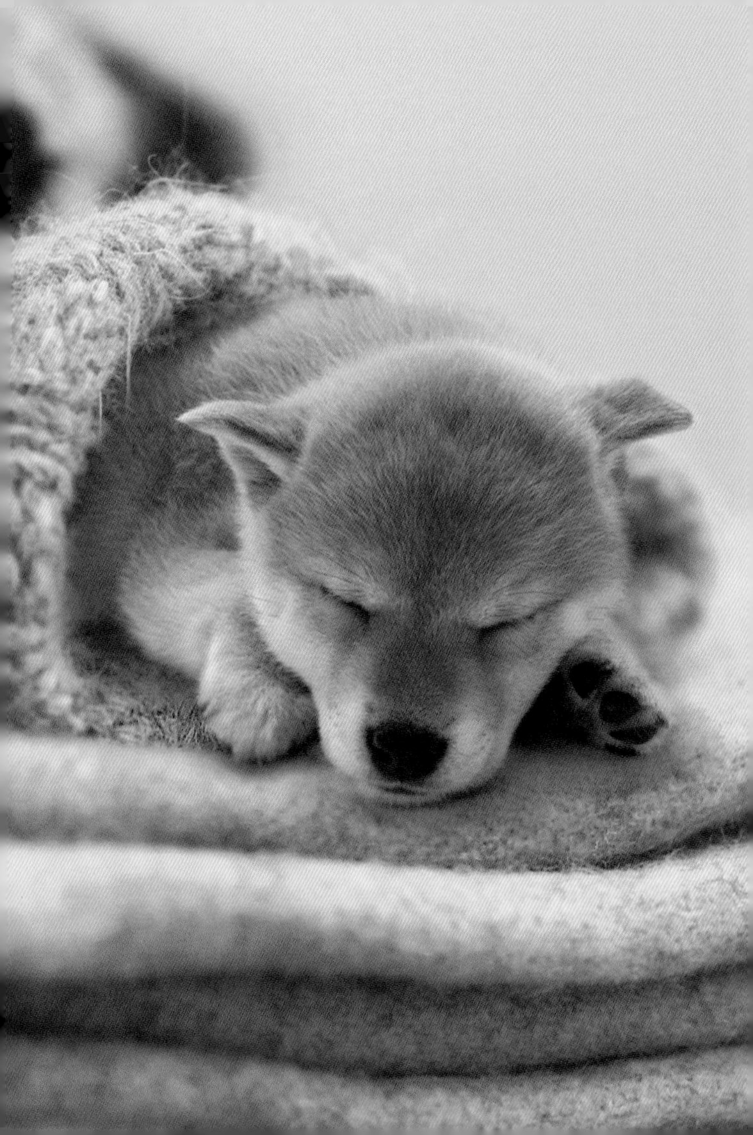

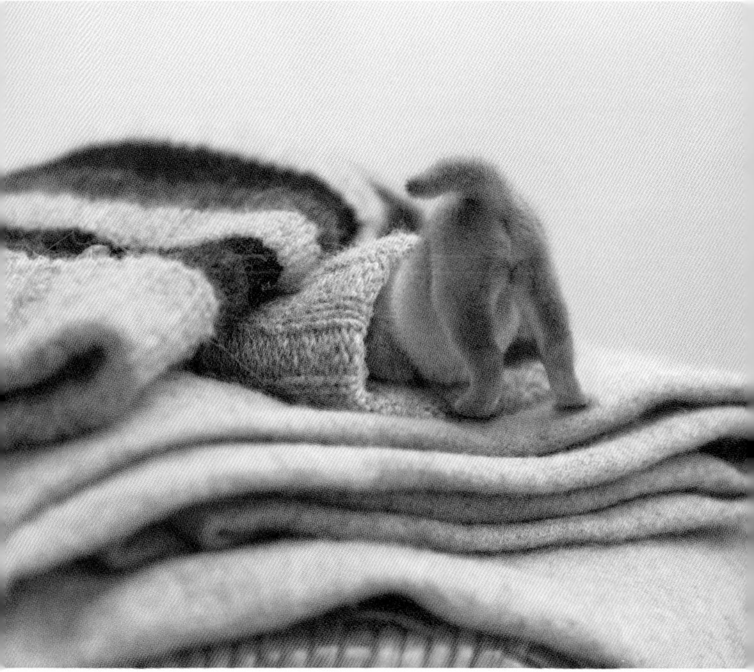

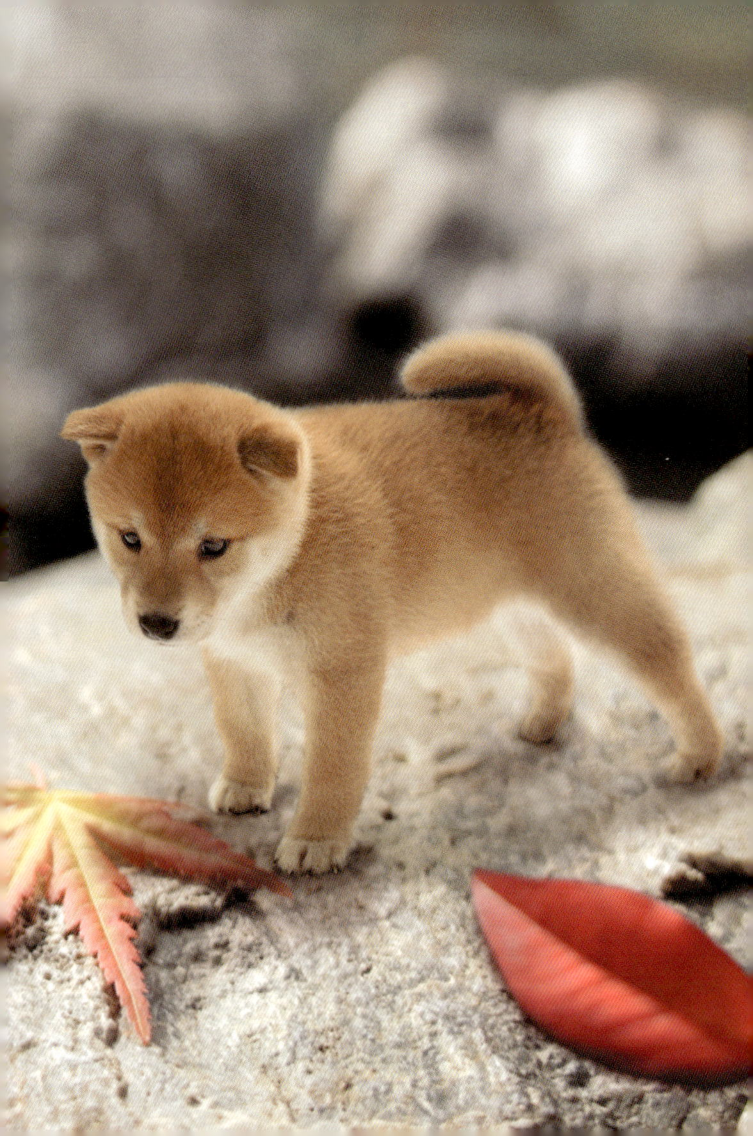

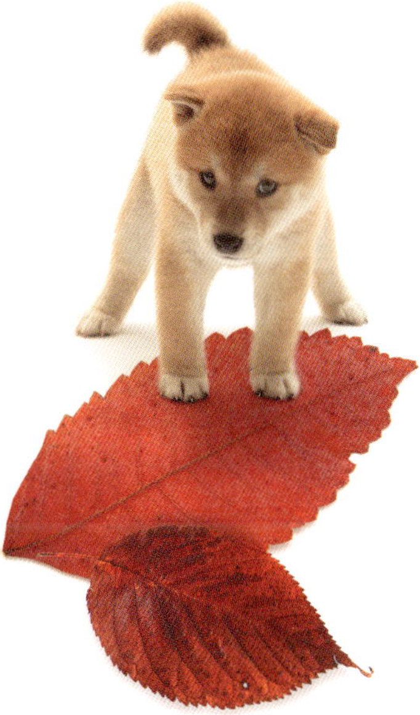

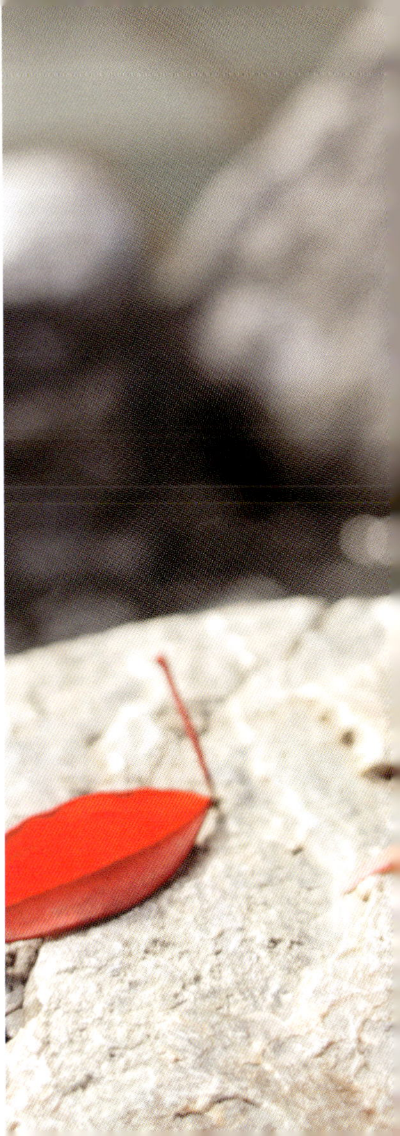

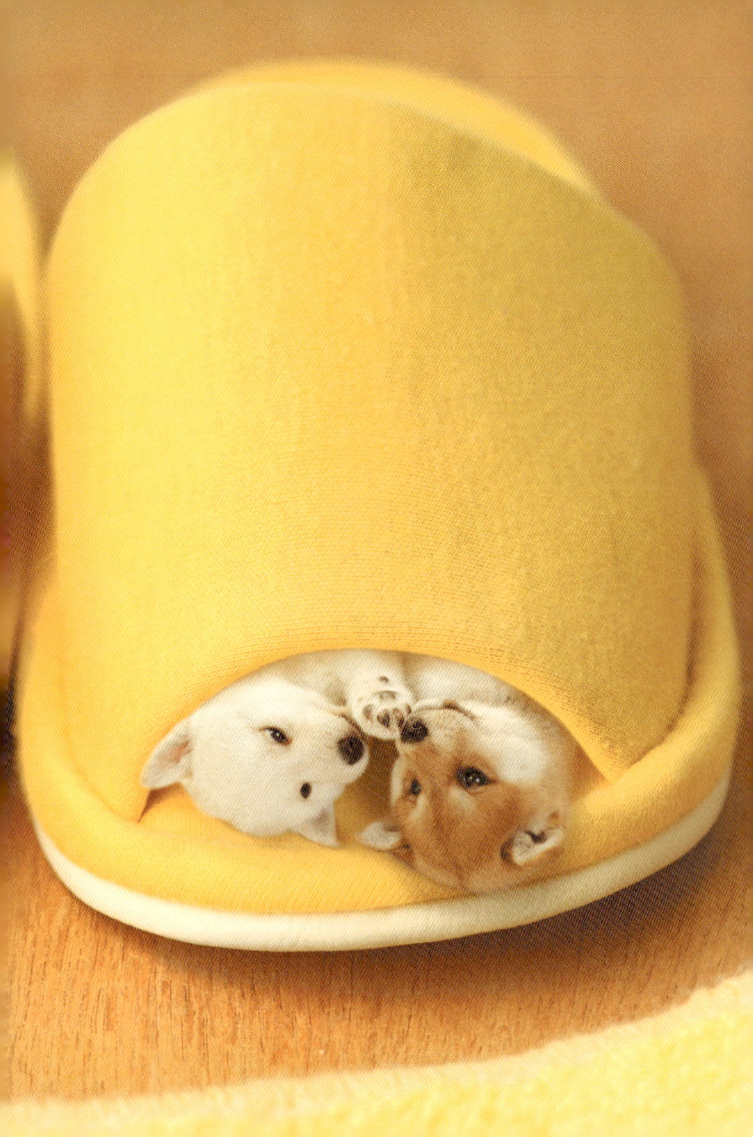

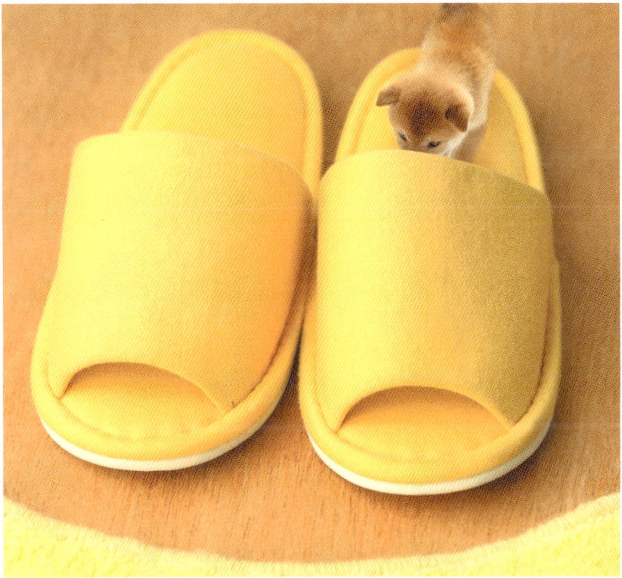

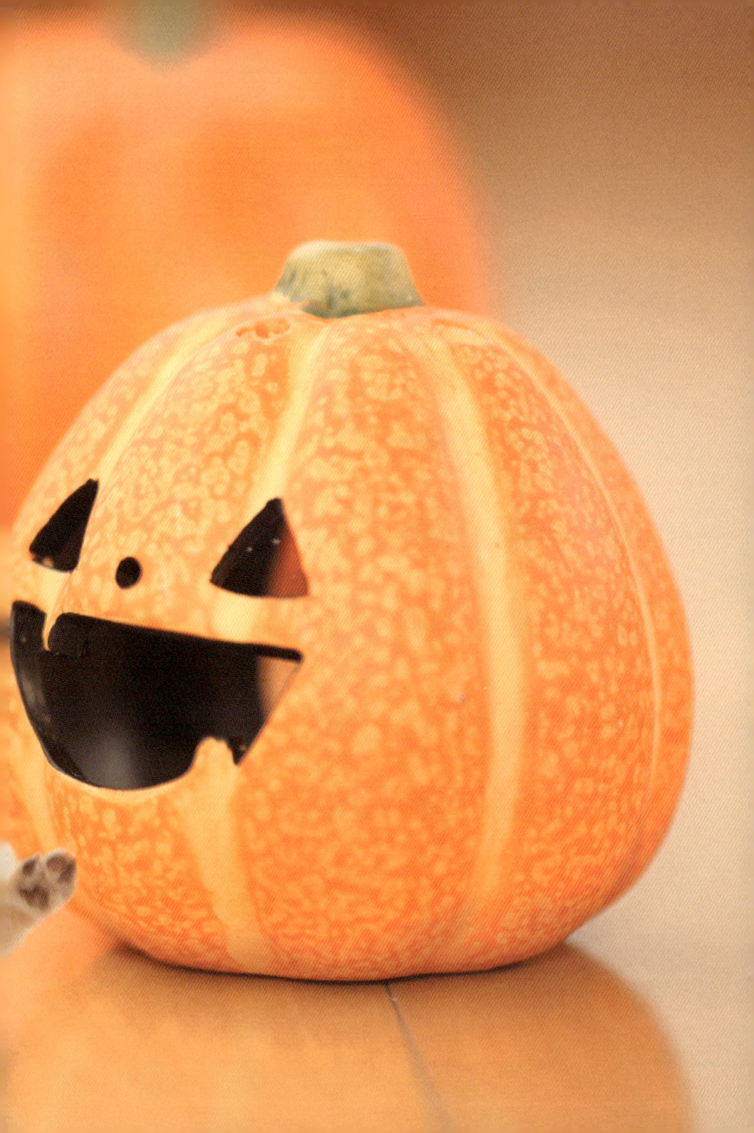

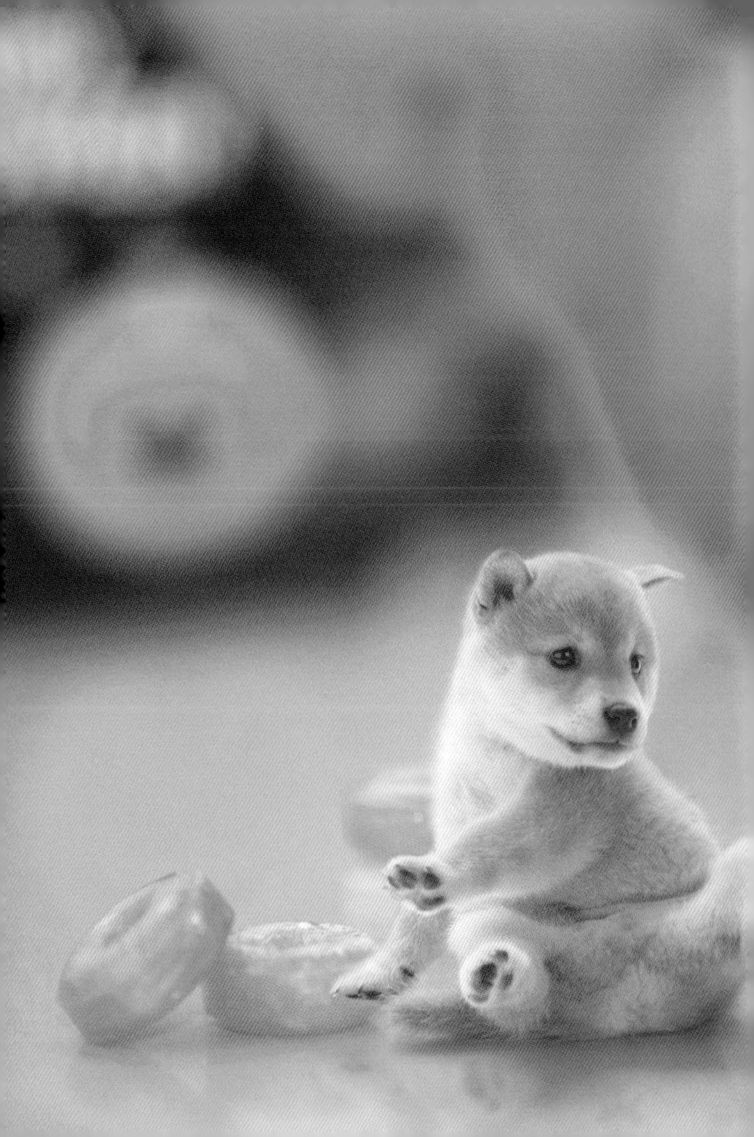

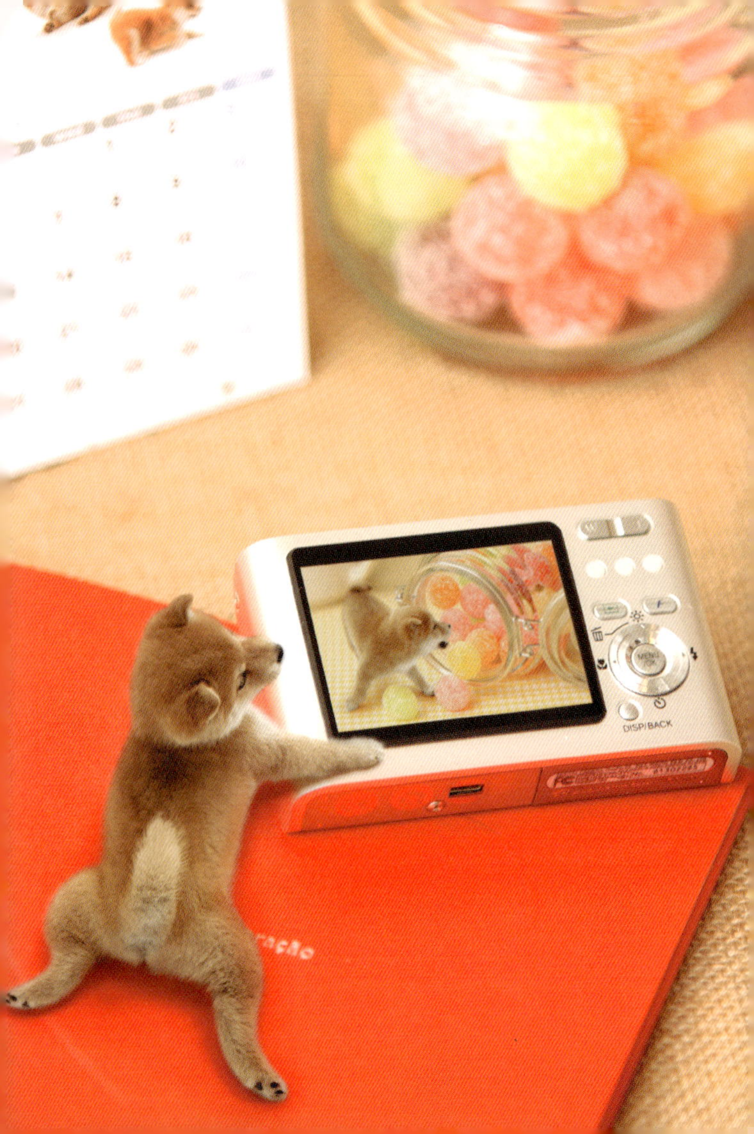
DISP/BACK

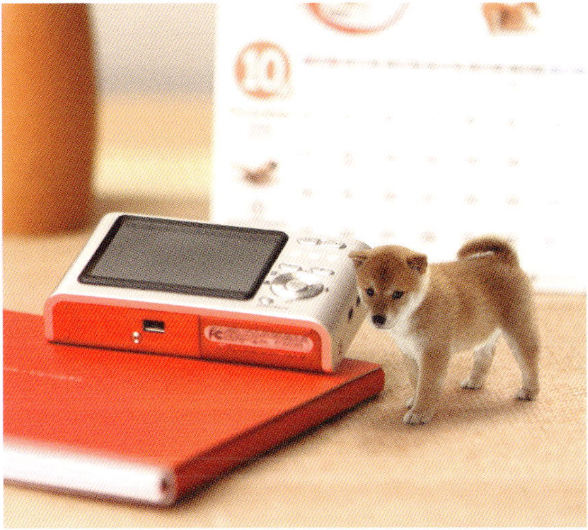

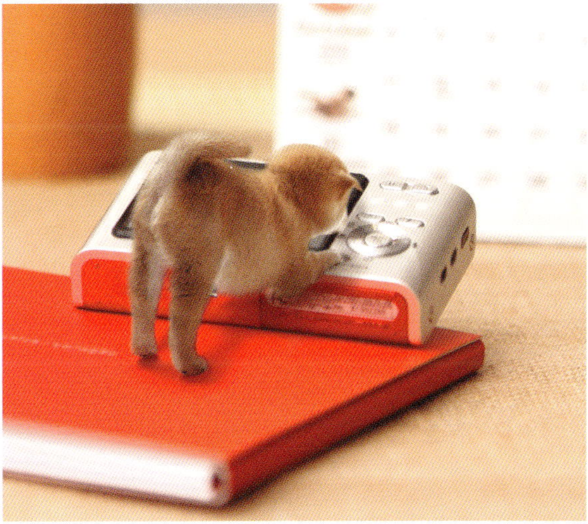

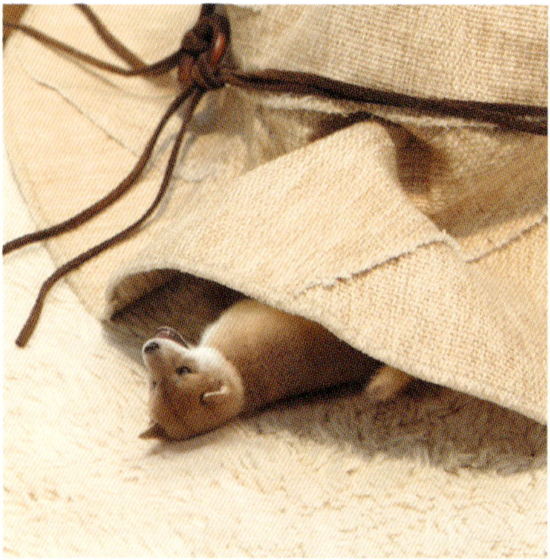

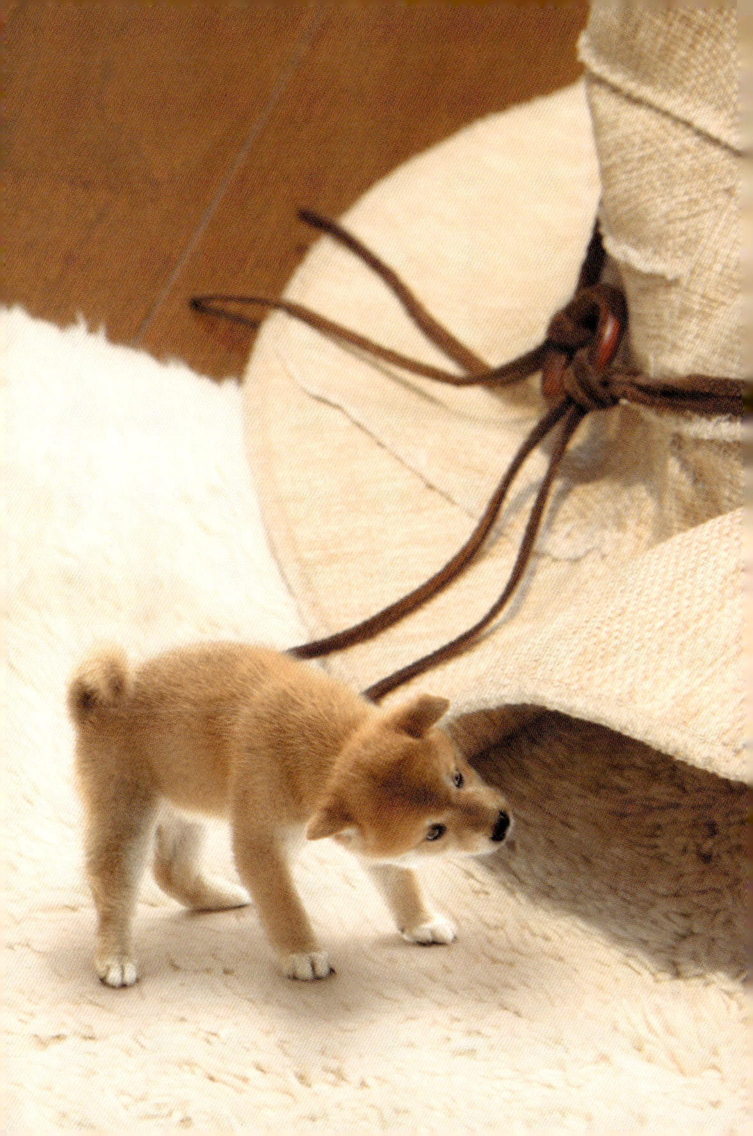

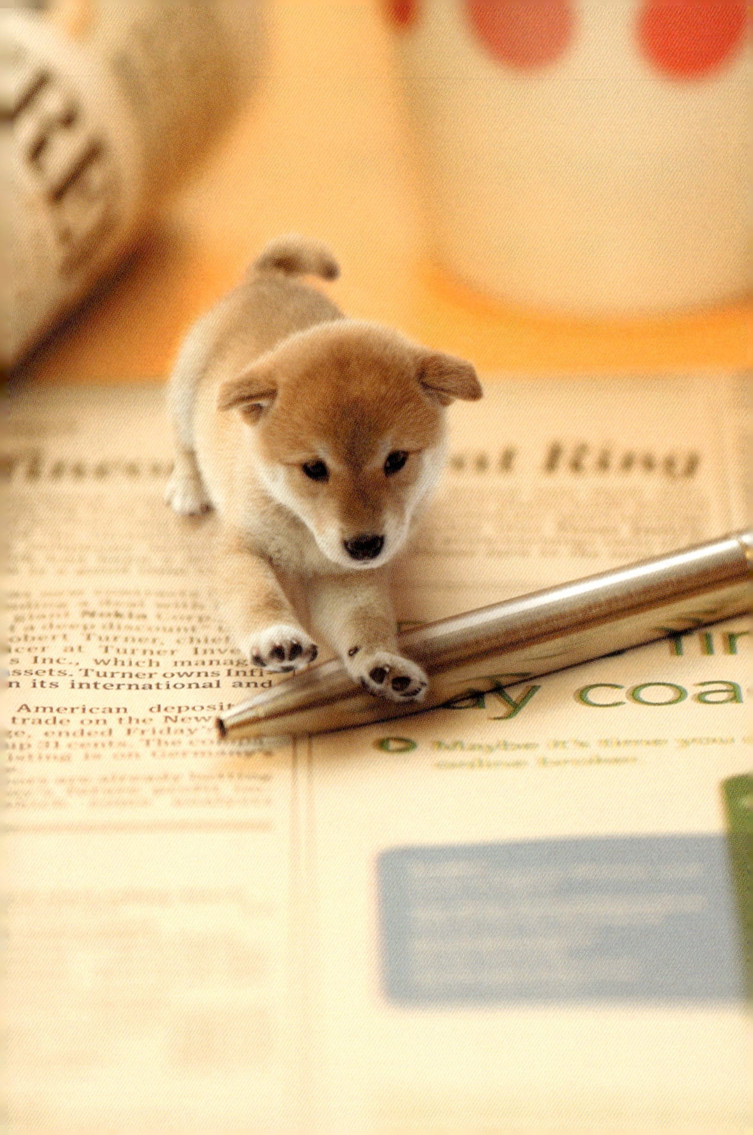
s Inc., which manag
ssets. Turner owns Infi-
n its international and
American deposit
trade on the New
coa

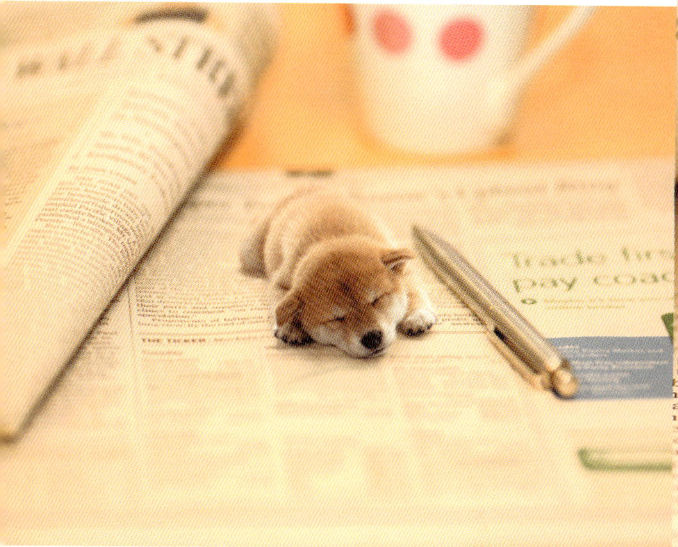

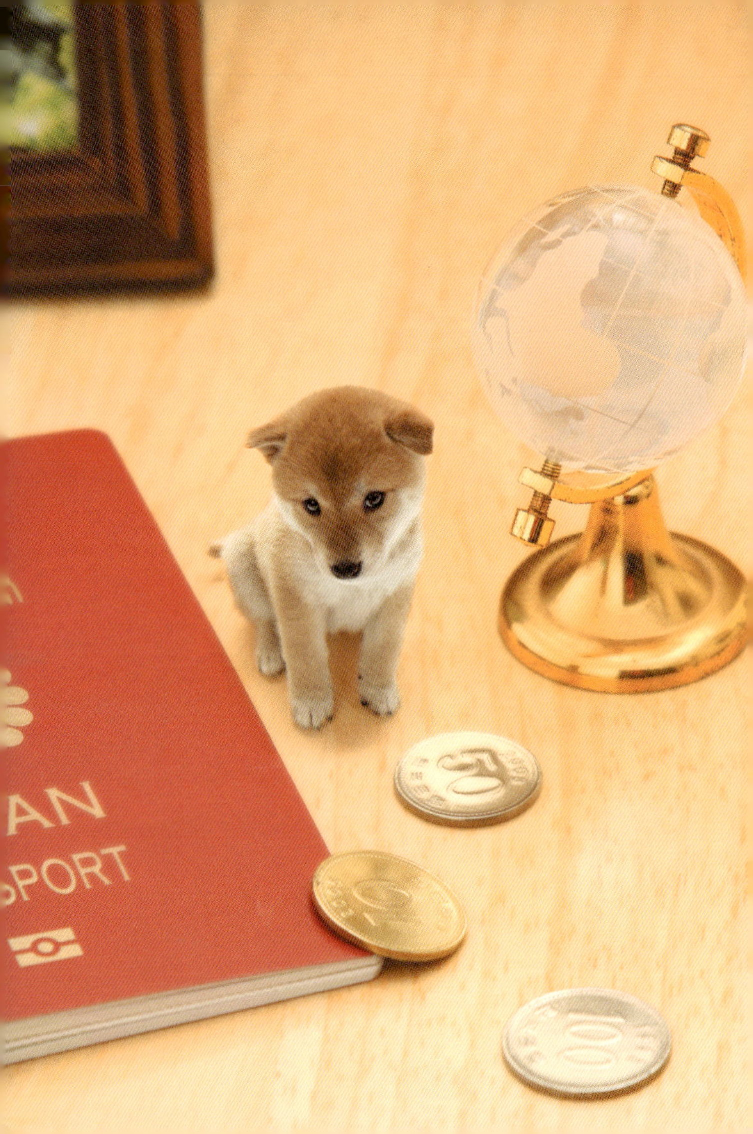
AN
SPORT

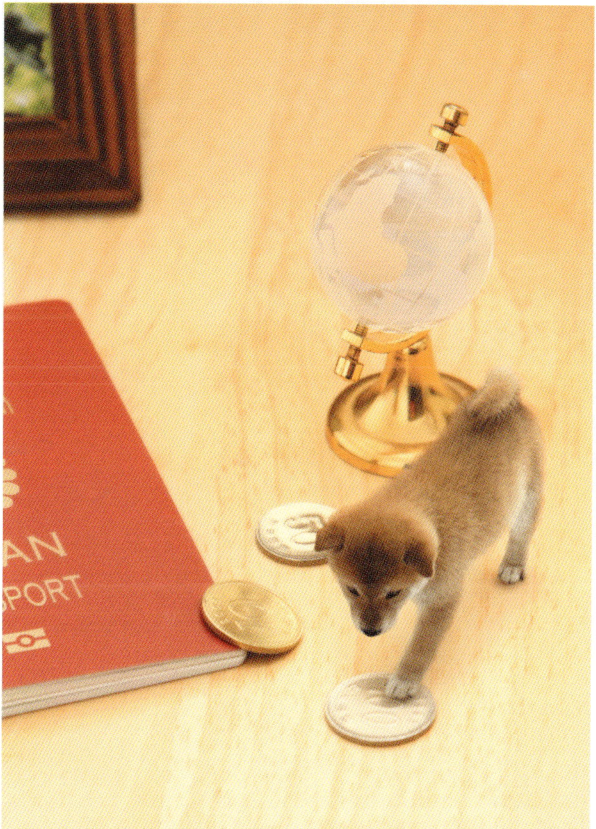

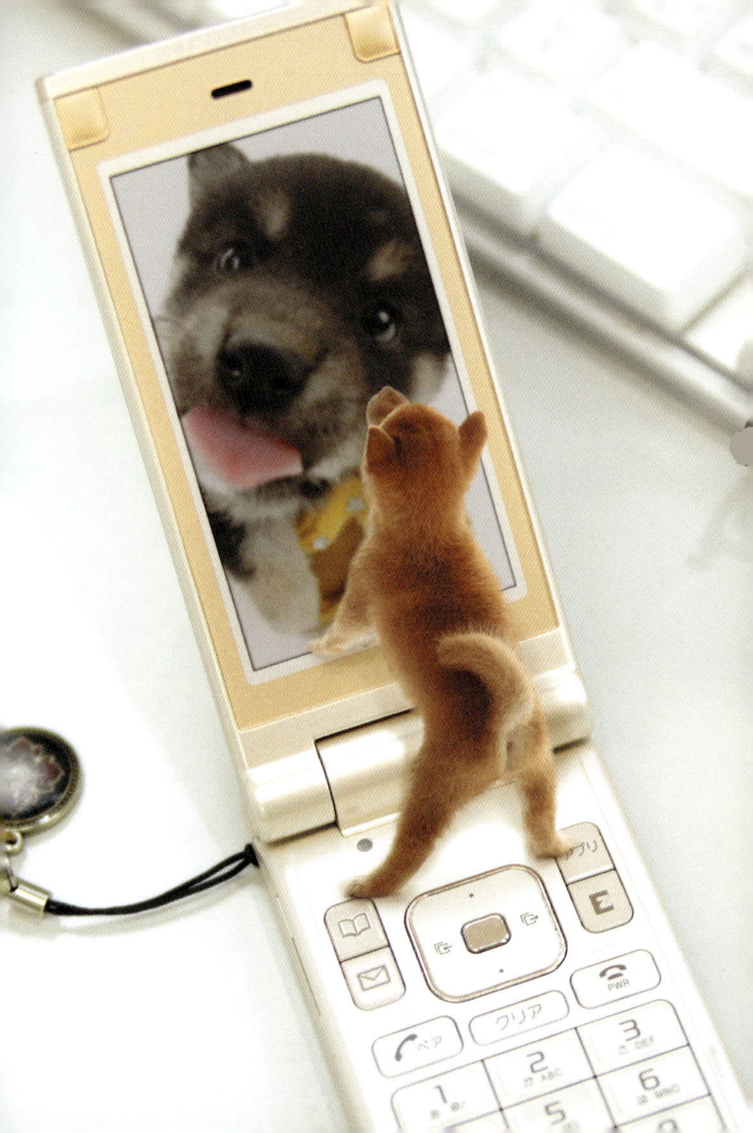
E
PWR
クリア
ペア
3
2
1
6
5

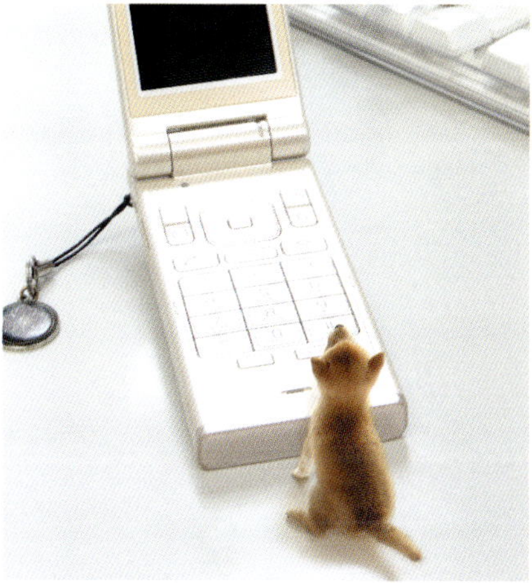

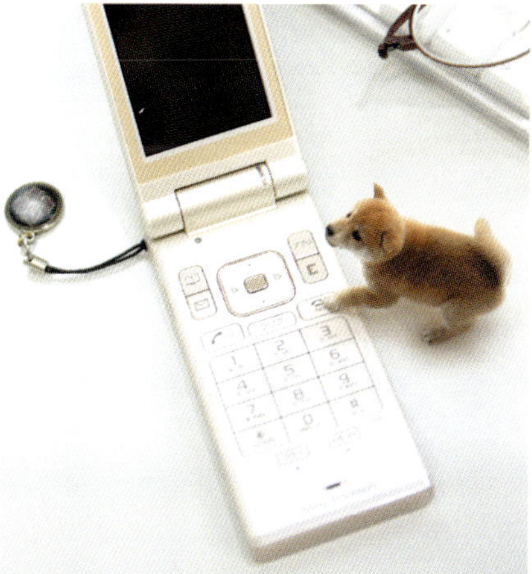

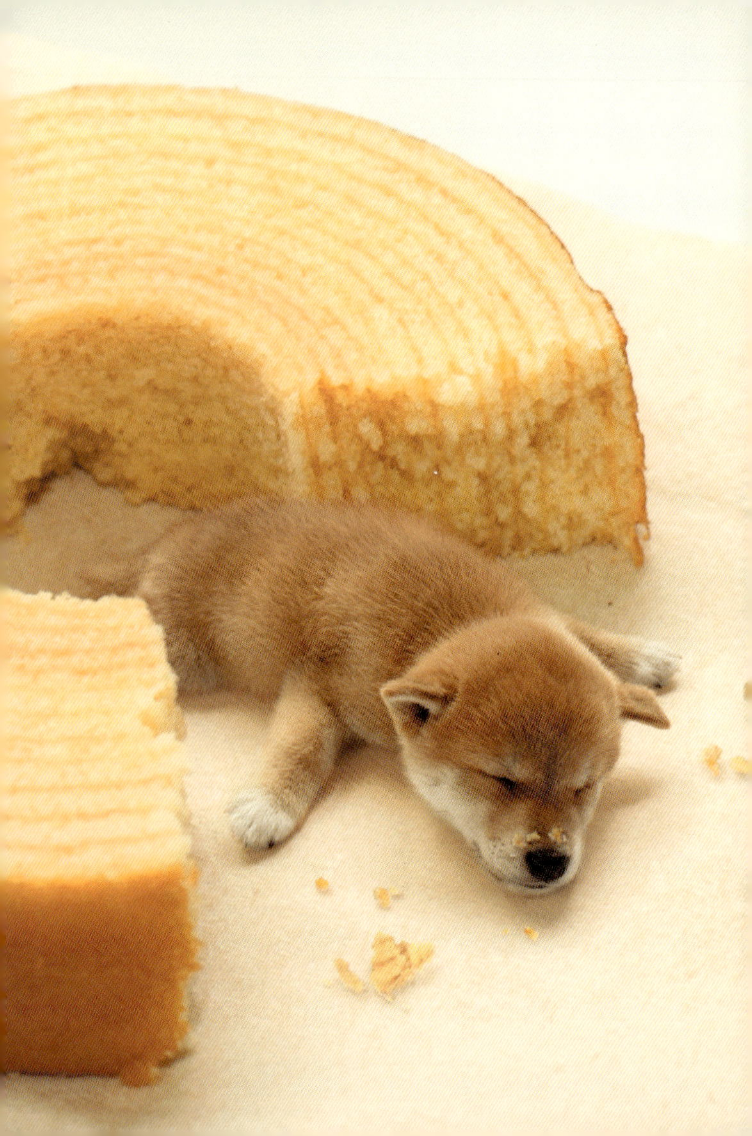

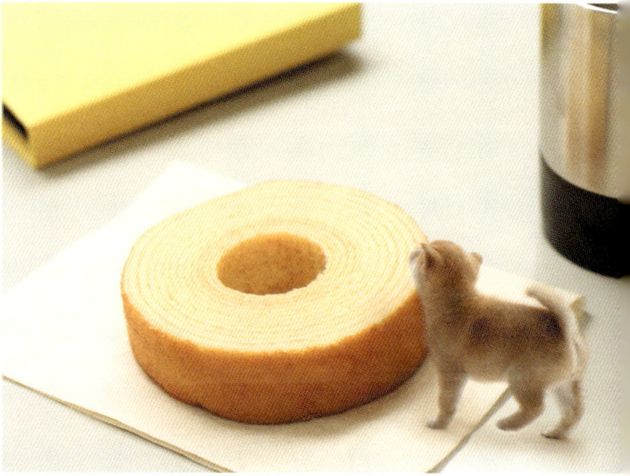

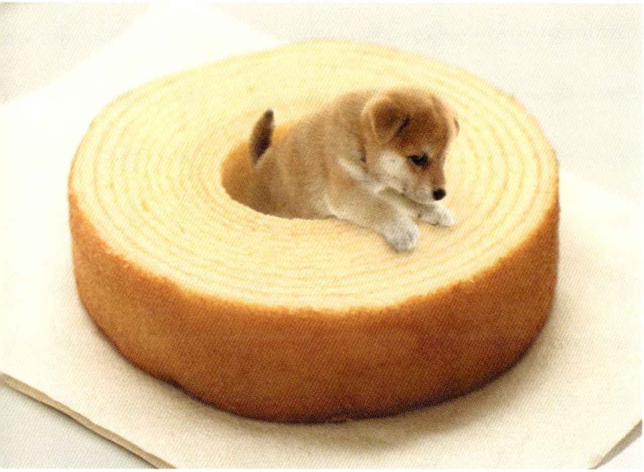

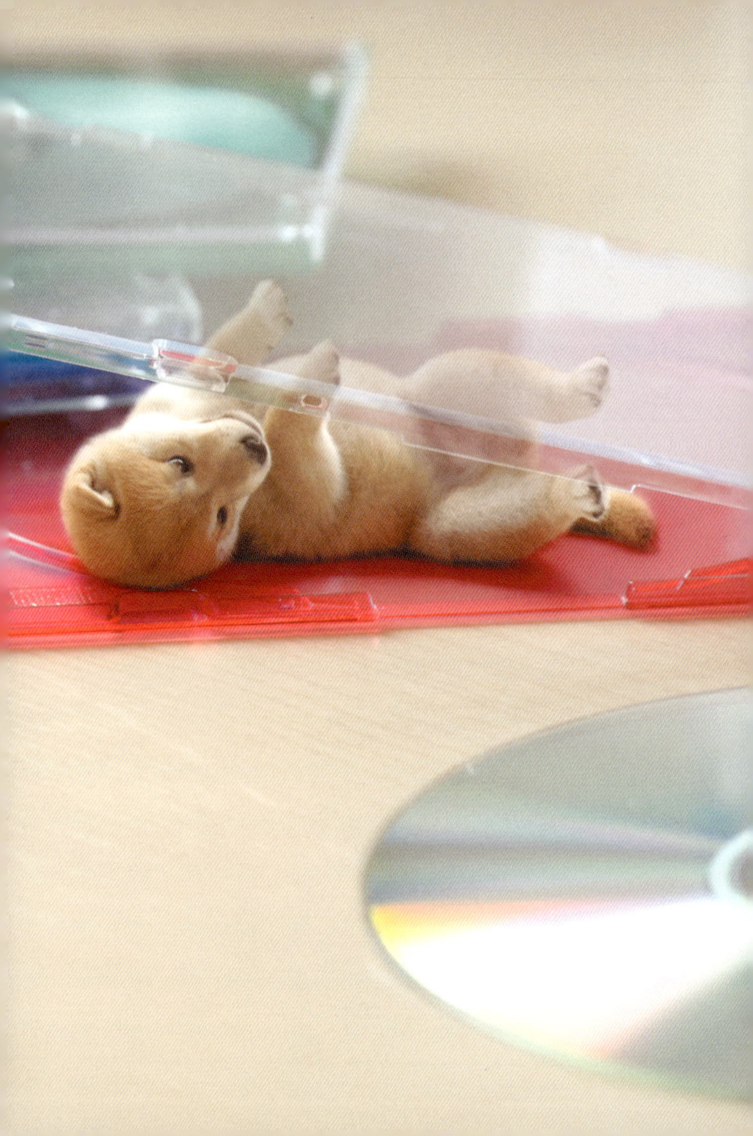

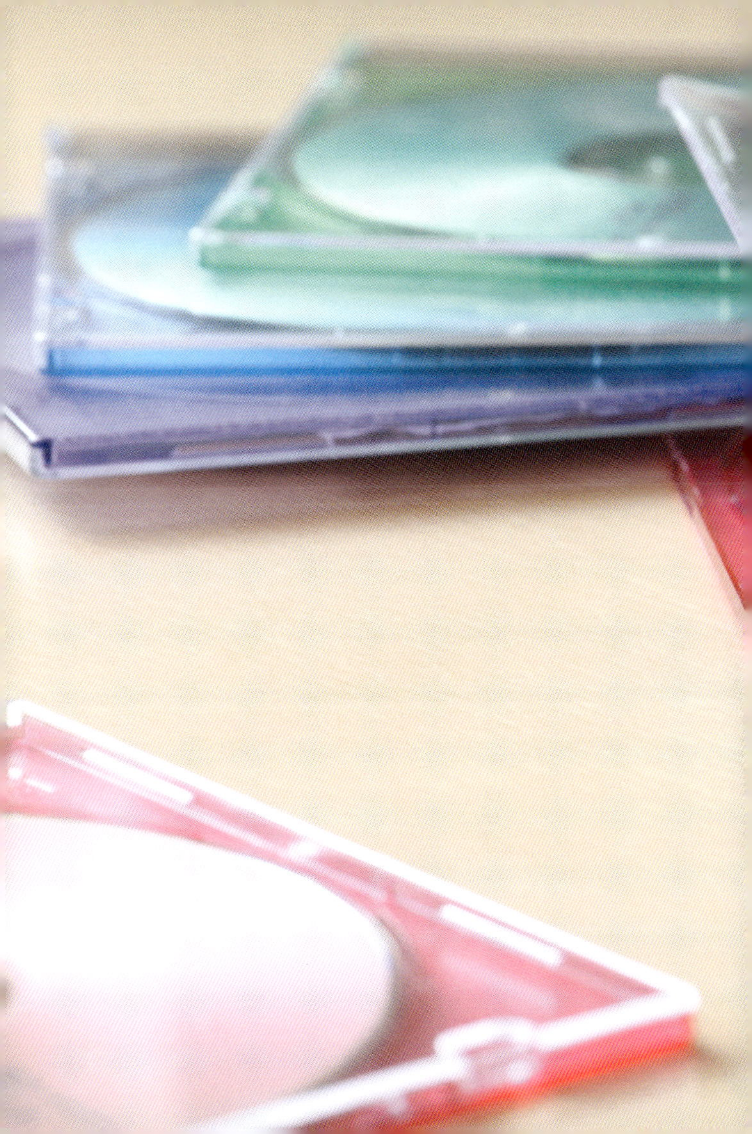

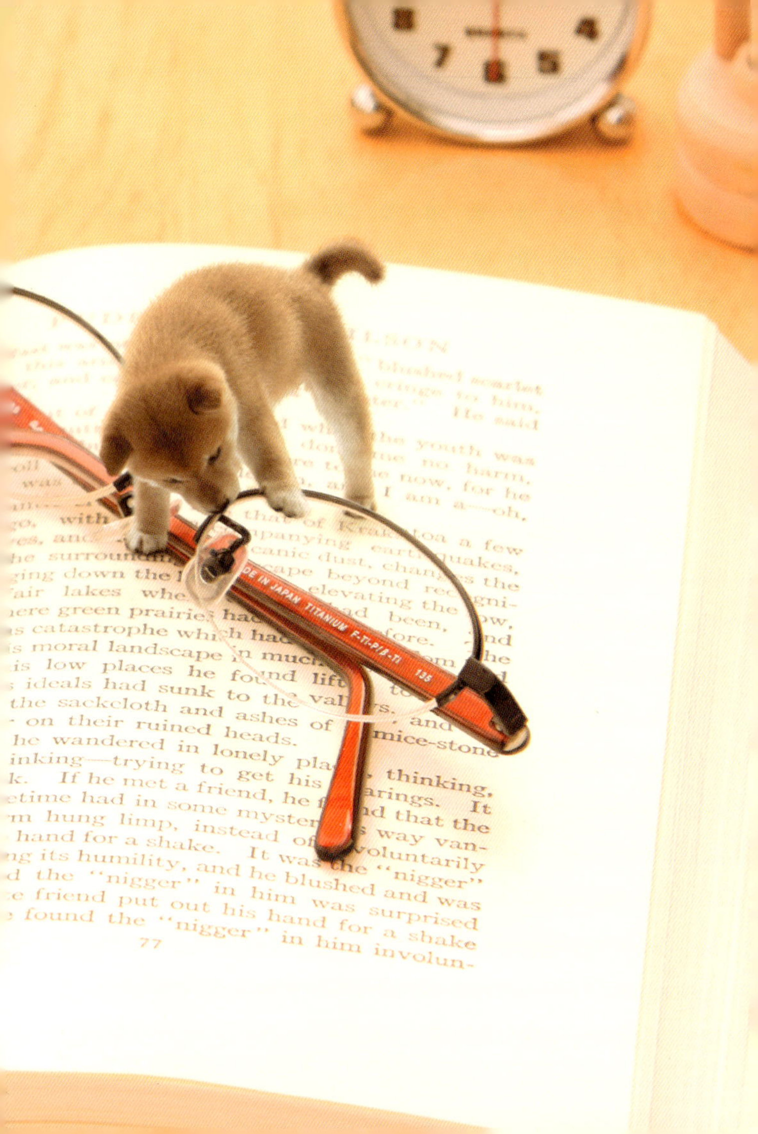

TITANIUM F-TI-P/β-TI
135
ng its humility, and he blushed the "nigger"
d the "nigger" in him blushed and was
e friend put out his hand was surprised
found the "nigger" in him involun-
77

CHAPTER X

THE NYMPH REVEALED

All say, "How hard it is that we have to die"—a strange complaint to come from the mouths of people who have had to live.

—*Pudd'nhead Wilson's Calendar.*

When angry, count four; when very angry, swear.

—*Pudd'nhead Wilson's Calendar.*

EVERY now and then, after Tom went to bed, he had sudden wakings out of his sleep, and his first thought was, "Oh, joy, it was all a dream!" Then he laid himself heavily down again, with a groan and the muttered words, "A nigger! I am a nigger! Oh, I wish I was dead!"

He woke at dawn with one more repetition of this horror, and then he resolved to meddle no more with that treacherous sleep. He began to think. Sufficiently bitter thinkings they were. They wandered along something after this fashion:

"Why were niggers *and* whites made? What crime did the uncreated first nigger commit that the curse of birth was decreed for him? And why is this awful difference made between white and black? . . . How hard the nigger's fate seems, this morning!—yet until last night such a thought never entered my head."

He sighed and groaned an hour or more away. Then "Chambers" came humbly in to say that

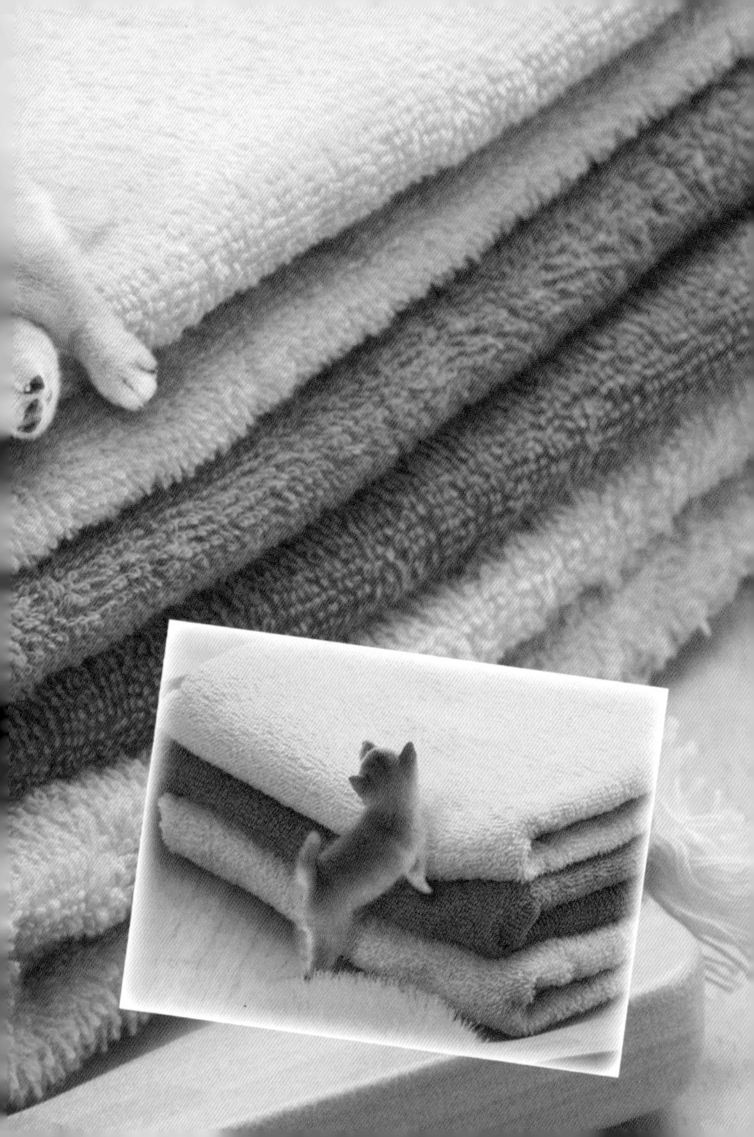

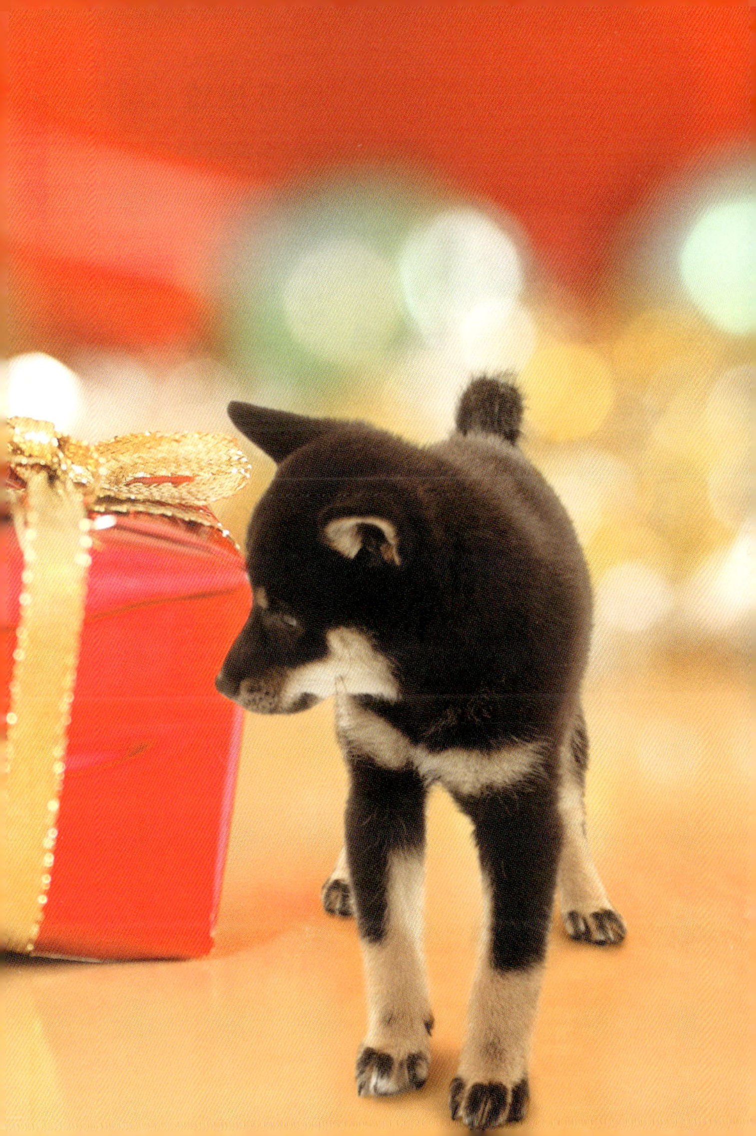

POST CARD

きりとり線

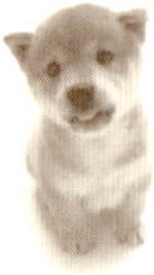